Forgotten Books

RELATIVITY

THE SPECIAL & THE GENERAL THEORY

A POPULAR EXPOSITION

BY

ALBERT EINSTEIN, Ph.D.

PROFESSOR OF PHYSICS IN THE UNIVERSITY OF BERLIN

AUTHORISED TRANSLATION BY

ROBERT W. LAWSON, D.Sc.

UNIVERSITY OF SHEFFIELD

WITH FIVE DIAGRAMS
AND A PORTRAIT OF THE AUTHOR

THIRD EDITION

This Translation was first Published . . August 19th 1920
Second Edition . . . . September 1920
Third Edition 1920

PREFACE

THE present book is intended, as far as possible, to give an exact insight into the theory of Relativity to those readers who, from a general scientific and philosophical point of view, are interested in the theory, but who are not conversant with the mathematical apparatus[1] of theoretical physics. The work presumes a standard of education corresponding to that of a university matriculation examination, and, despite the shortness of the book, a fair amount of patience and force of will on the part of the reader. The author has spared himself no pains in his endeavour

[1] The mathematical fundaments of the special theory of relativity are to be found in the original papers of H. A. Lorentz, A. Einstein, H. Minkowski, published under the title *Das Relativitätsprinzip* (The Principle of Relativity) in B. G. Teubner's collection of monographs *Fortschritte der mathematischen Wissenschaften* (Advances in the Mathematical Sciences), also in M. Laue's exhaustive book *Das Relativitätsprinzip*—published by Friedr. Vieweg & Son, Braunschweig. The general theory of relativity, together with the necessary parts of the theory of invariants, is dealt with in the author's book *Die Grundlagen der allgemeinen Relativitätstheorie* (The Foundations of the General Theory of Relativity)—Joh. Ambr. Barth, 1916; this book assumes some familiarity with the special theory of relativity.

to present the main ideas in the simplest and most intelligible form, and on the whole, in the sequence and connection in which they actually originated. In the interest of clearness, it appeared to me inevitable that I should repeat myself frequently, without paying the slightest attention to the elegance of the presentation. I adhered scrupulously to the precept of that brilliant theoretical physicist L. Boltzmann, according to whom matters of elegance ought to be left to the tailor and to the cobbler. I make no pretence of having withheld from the reader difficulties which are inherent to the subject. On the other hand, I have purposely treated the empirical physical foundations of the theory in a "step-motherly" fashion, so that readers unfamiliar with physics may not feel like the wanderer who was unable to see the forest for trees. May the book bring some one a few happy hours of suggestive thought !

December, 1916 A. EINSTEIN

NOTE TO THE THIRD EDITION

IN the present year (1918) an excellent and detailed manual on the general theory of relativity, written by H. Weyl, was published by the firm Julius Springer (Berlin). This book, entitled *Raum—Zeit—Materie* (Space—Time—Matter), may be warmly recommended to mathematicians and physicists.

BIOGRAPHICAL NOTE

ALBERT EINSTEIN is the son of German-Jewish parents. He was born in 1879 in the town of Ulm, Würtemberg, Germany. His schooldays were spent in Munich, where he attended the *Gymnasium* until his sixteenth year. After leaving school at Munich, he accompanied his parents to Milan, whence he proceeded to Switzerland six months later to continue his studies.

From 1896 to 1900 Albert Einstein studied mathematics and physics at the Technical High School in Zurich, as he intended becoming a secondary school (*Gymnasium*) teacher. For some time afterwards he was a private tutor, and having meanwhile become naturalised, he obtained a post as engineer in the Swiss Patent Office in 1902, which position he occupied till 1909. The main ideas involved in the most important of Einstein's theories date back to this period. Amongst these may be mentioned: *The Special Theory of Relativity, Inertia of Energy, Theory of the Brownian Movement*, and the *Quantum-Law of the Emission and Absorption of Light* (1905). These were followed some years later by the *Theory of the Specific Heat of Solid Bodies*, and the fundamental idea of the *General Theory of Relativity*.

During the interval 1909 to 1911 he occupied the post of Professor *Extraordinarius* at the University of Zurich, afterwards being appointed to the University of Prague, Bohemia, where he remained as Professor *Ordinarius* until 1912. In the latter year Professor Einstein accepted a similar chair at the *Polytechnikum*, Zurich, and continued his activities there until 1914, when he received a call to the Prussian Academy of Science, Berlin, as successor to Van't Hoff. Professor Einstein is able to devote himself freely to his studies at the Berlin Academy, and it was here that he succeeded in completing his work on the *General Theory of Relativity* (1915-17). Professor Einstein also lectures on various special branches of physics at the University of Berlin, and, in addition, he is Director of the Institute for Physical Research of the *Kaiser Wilhelm Gesellschaft*.

Professor Einstein has been twice married. His first wife, whom he married at Berne in 1903, was a fellow-student from Serbia. There were two sons of this marriage, both of whom are living in Zurich, the elder being sixteen years of age. Recently Professor Einstein married a widowed cousin, with whom he is now living in Berlin.

R. W. L.

TRANSLATOR'S NOTE

IN presenting this translation to the English-reading public, it is hardly necessary for me to enlarge on the Author's prefatory remarks, except to draw attention to those additions to the book which do not appear in the original.

At my request, Professor Einstein kindly supplied me with a portrait of himself, by one of Germany's most celebrated artists. Appendix III, on "The Experimental Confirmation of the General Theory of Relativity," has been written specially for this translation. Apart from these valuable additions to the book, I have included a biographical note on the Author, and, at the end of the book, an Index and a list of English references to the subject. This list, which is more suggestive than exhaustive, is intended as a guide to those readers who wish to pursue the subject farther.

I desire to tender my best thanks to my colleagues Professor S. R. Milner, D.Sc., and Mr. W. E. Curtis, A.R.C.Sc., F.R.A.S., also to my friend Dr. Arthur Holmes, A.R.C.Sc., F.G.S., of the Imperial College, for their kindness in reading through the manuscript, for helpful criticism, and for numerous suggestions. I owe an expression of thanks also to Messrs. Methuen for their ready counsel and advice, and for the care they have bestowed on the work during the course of its publication.

ROBERT W. LAWSON

THE PHYSICS LABORATORY
THE UNIVERSITY OF SHEFFIELD
June 12, 1920

CONTENTS

PART I

THE SPECIAL THEORY OF RELATIVITY

PART II

THE GENERAL THEORY OF RELATIVITY

PART III

CONSIDERATIONS ON THE UNIVERSE AS A WHOLE

APPENDICES

RELATIVITY

THE SPECIAL AND THE GENERAL THEORY

RELATIVITY

PART I

THE SPECIAL THEORY OF RELATIVITY

I

PHYSICAL MEANING OF GEOMETRICAL PROPOSITIONS

IN your schooldays most of you who read this book made acquaintance with the noble building of Euclid's geometry, and you remember—perhaps with more respect than love—the magnificent structure, on the lofty staircase of which you were chased about for uncounted hours by conscientious teachers. By reason of your past experience, you would certainly regard everyone with disdain who should pronounce even the most out-of-the-way proposition of this science to be untrue. But perhaps this feeling of proud certainty would leave you immediately if some one were to ask you: "What, then, do you mean by the assertion that these propositions are true?" Let us proceed to give this question a little consideration.

Geometry sets out from certain conceptions such as "plane," "point," and "straight line," with which we are able to associate more or less definite ideas, and from certain simple propositions (axioms) which, in virtue of these ideas, we are inclined to accept as "true." Then, on the basis of a logical process, the justification of which we feel ourselves compelled to admit, all remaining propositions are shown to follow from those axioms, *i.e.* they are proven. A proposition is then correct ("true") when it has been derived in the recognised manner from the axioms. The question of the "truth" of the individual geometrical propositions is thus reduced to one of the "truth" of the axioms. Now it has long been known that the last question is not only unanswerable by the methods of geometry, but that it is in itself entirely without meaning. We cannot ask whether it is true that only one straight line goes through two points. We can only say that Euclidean geometry deals with things called "straight lines," to each of which is ascribed the property of being uniquely determined by two points situated on it. The concept "true" does not tally with the assertions of pure geometry, because by the word "true" we are eventually in the habit of designating always the correspondence with a "real" object; geometry, however, is not concerned with the relation of the ideas involved in it to objects of experience, but only with the logical connection of these ideas among themselves.

It is not difficult to understand why, in spite of this, we feel constrained to call the propositions of geometry "true." Geometrical ideas correspond to more or less exact objects in nature, and these last are undoubtedly the exclusive cause of the genesis of those ideas. Geometry ought to refrain from such a course, in order to give to its structure the largest possible logical unity. The practice, for example, of seeing in a "distance" two marked positions on a practically rigid body is something which is lodged deeply in our habit of thought. We are accustomed further to regard three points as being situated on a straight line, if their apparent positions can be made to coincide for observation with one eye, under suitable choice of our place of observation.

If, in pursuance of our habit of thought, we now supplement the propositions of Euclidean geometry by the single proposition that two points on a practically rigid body always correspond to the same distance (line-interval), independently of any changes in position to which we may subject the body, the propositions of Euclidean geometry then resolve themselves into propositions on the possible relative position of practically rigid bodies.[1] Geometry which has been supplemented in this way is then to be treated as a branch of physics. We can now legitimately ask as to the "truth" of geometrical propositions interpreted in this way, since we are justified in asking whether these propositions are satisfied for those real things we have associated with the geometrical ideas. In less exact terms we can express this by saying that by the "truth" of a geometrical proposition in this sense we understand its validity for a construction with ruler and compasses.

[1] It follows that a natural object is associated also with a straight line. Three points A, B and C on a rigid body thus lie in a straight line when, the points A and C being given, B is chosen such that the sum of the distances AB and BC is as short as possible. This incomplete suggestion will suffice for our present purpose.

Of course the conviction of the "truth" of geometrical propositions in this sense is founded exclusively on rather incomplete experience. For the present we shall assume the "truth" of the geometrical propositions, then at a later stage (in the general theory of relativity) we shall see that this "truth" is limited, and we shall consider the extent of its limitation.

II

THE SYSTEM OF CO-ORDINATES

ON the basis of the physical interpretation of distance which has been indicated, we are also in a position to establish the distance between two points on a rigid body by means of measurements. For this purpose we require a "distance" (rod *S*) which is to be used once and for all, and which we employ as a standard measure. If, now, *A* and *B* are two points on a rigid body, we can construct the line joining them according to the rules of geometry; then, starting from *A*, we can mark off the distance *S* time after time until we reach *B*. The number of these operations required is the numerical measure of the distance *AB*. This is the basis of all measurement of length.[1]

Every description of the scene of an event or of the position of an object in space is based on the specification of the point on a rigid body (body of reference) with which that event or object coincides. This applies not only to scientific description, but also to everyday life. If I analyse the place specification "Trafalgar

[1] Here we have assumed that there is nothing left over, *i.e.* that the measurement gives a whole number. This difficulty is got over by the use of divided measuring-rods, the introduction of which does not demand any fundamentally new method.

Square, London,"[1] I arrive at the following result. The earth is the rigid body to which the specification of place refers; "Trafalgar Square, London," is a well-defined point, to which a name has been assigned, and with which the event coincides in space.[2]

This primitive method of place specification deals only with places on the surface of rigid bodies, and is dependent on the existence of points on this surface which are distinguishable from each other. But we can free ourselves from both of these limitations without altering the nature of our specification of position. If, for instance, a cloud is hovering over Trafalgar Square, then we can determine its position relative to the surface of the earth by erecting a pole perpendicularly on the Square, so that it reaches the cloud. The length of the pole measured with the standard measuring-rod, combined with the specification of the position of the foot of the pole, supplies us with a complete place specification. On the basis of this illustration, we are able to see the manner in which a refinement of the conception of position has been developed.

(*a*) We imagine the rigid body, to which the place specification is referred, supplemented in such a manner that the object whose position we require is reached by the completed rigid body.

(*b*) In locating the position of the object, we make use of a number (here the length of the pole measured

[1] I have chosen this as being more familiar to the English reader than the "Potsdamer Platz, Berlin," which is referred to in the original. (R. W. L.)

[2] It is not necessary here to investigate further the significance of the expression "coincidence in space." This conception is sufficiently obvious to ensure that differences of opinion are scarcely likely to arise as to its applicability in practice.

with the measuring-rod) instead of designated points of reference.

(c) We speak of the height of the cloud even when the pole which reaches the cloud has not been erected. By means of optical observations of the cloud from different positions on the ground, and taking into account the properties of the propagation of light, we determine the length of the pole we should have required in order to reach the cloud.

From this consideration we see that it will be advantageous if, in the description of position, it should be possible by means of numerical measures to make ourselves independent of the existence of marked positions (possessing names) on the rigid body of reference. In the physics of measurement this is attained by the application of the Cartesian system of co-ordinates.

This consists of three plane surfaces perpendicular to each other and rigidly attached to a rigid body. Referred to a system of co-ordinates, the scene of any event will be determined (for the main part) by the specification of the lengths of the three perpendiculars or co-ordinates (x, y, z) which can be dropped from the scene of the event to those three plane surfaces. The lengths of these three perpendiculars can be determined by a series of manipulations with rigid measuring-rods performed according to the rules and methods laid down by Euclidean geometry.

In practice, the rigid surfaces which constitute the system of co-ordinates are generally not available; furthermore, the magnitudes of the co-ordinates are not actually determined by constructions with rigid rods, but by indirect means. If the results of physics and astronomy are to maintain their clearness, the physical meaning of specifications of position must always be sought in accordance with the above considerations.[1]

We thus obtain the following result: Every description of events in space involves the use of a rigid body to which such events have to be referred. The resulting relationship takes for granted that the laws of Euclidean geometry hold for "distances," the "distance" being represented physically by means of the convention of two marks on a rigid body.

[1] A refinement and modification of these views does not become necessary until we come to deal with the general theory of relativity, treated in the second part of this book.

III

SPACE AND TIME IN CLASSICAL MECHANICS

"THE purpose of mechanics is to describe how bodies change their position in space with time." I should load my conscience with grave sins against the sacred spirit of lucidity were I to formulate the aims of mechanics in this way, without serious reflection and detailed explanations. Let us proceed to disclose these sins.

It is not clear what is to be understood here by "position" and "space." I stand at the window of a railway carriage which is travelling uniformly, and drop a stone on the embankment, without throwing it. Then, disregarding the influence of the air resistance, I see the stone descend in a straight line. A pedestrian who observes the misdeed from the footpath notices that the stone falls to earth in a parabolic curve. I now ask: Do the "positions" traversed by the stone lie "in reality" on a straight line or on a parabola? Moreover, what is meant here by motion "in space"? From the considerations of the previous section the answer is self-evident. In the first place, we entirely shun the vague word "space," of which, we must honestly acknowledge, we cannot form the slightest conception, and we replace it by "motion relative to a practically rigid body of reference." The positions relative to the body of reference (railway carriage or embankment) have already been defined in detail in the preceding section. If instead of " body of reference " we insert " system of co-ordinates," which is a useful idea for mathematical description, we are in a position to say: The stone traverses a straight line relative to a system of co-ordinates rigidly attached to the carriage, but relative to a system of co-ordinates rigidly attached to the ground (embankment) it describes a parabola. With the aid of this example it is clearly seen that there is no such thing as an independently existing trajectory (lit. " path-curve " [1]), but only a trajectory relative to a particular body of reference.

In order to have a *complete* description of the motion, we must specify how the body alters its position *with time*; *i.e.* for every point on the trajectory it must be stated at what time the body is situated there. These data must be supplemented by such a definition of time that, in virtue of this definition, these time-values can be regarded essentially as magnitudes (results of measurements) capable of observation. If we take our stand on the ground of classical mechanics, we can satisfy this requirement for our illustration in the following manner. We imagine two clocks of identical construction; the man at the railway-carriage window is holding one of them, and the man on the foot-path the other. Each of the observers determines the position on his own reference-body occupied by the stone at each tick of the clock he is holding in his hand. In this connection we have not taken account of the inaccuracy involved by the finiteness of the velocity of propagation of light. With this and with a second difficulty prevailing here we shall have to deal in detail later.

[1] That is, a curve along which the body moves.

IV

THE GALILEIAN SYSTEM OF CO-ORDINATES

AS is well known, the fundamental law of the mechanics of Galilei-Newton, which is known as the *law of inertia*, can be stated thus: A body removed sufficiently far from other bodies continues in a state of rest or of uniform motion in a straight line. This law not only says something about the motion of the bodies, but it also indicates the reference-bodies or systems of co-ordinates, permissible in mechanics, which can be used in mechanical description. The visible fixed stars are bodies for which the law of inertia certainly holds to a high degree of approximation. Now if we use a system of co-ordinates which is rigidly attached to the earth, then, relative to this system, every fixed star describes a circle of immense radius in the course of an astronomical day, a result which is opposed to the statement of the law of inertia. So that if we adhere to this law we must refer these motions only to systems of co-ordinates relative to which the fixed stars do not move in a circle. A system of co-ordinates of which the state of motion is such that the law of inertia holds relative to it is called a " Galileian system of co-ordinates." The laws of the mechanics of Galilei-Newton can be regarded as valid only for a Galileian system of co-ordinates.

V

THE PRINCIPLE OF RELATIVITY (IN THE RESTRICTED SENSE)

IN order to attain the greatest possible clearness, let us return to our example of the railway carriage supposed to be travelling uniformly. We call its motion a uniform translation ("uniform" because it is of constant velocity and direction, "translation" because although the carriage changes its position relative to the embankment yet it does not rotate in so doing). Let us imagine a raven flying through the air in such a manner that its motion, as observed from the embankment, is uniform and in a straight line. If we were to observe the flying raven from the moving railway carriage, we should find that the motion of the raven would be one of different velocity and direction, but that it would still be uniform and in a straight line. Expressed in an abstract manner we may say: If a mass m is moving uniformly in a straight line with respect to a co-ordinate system K, then it will also be moving uniformly and in a straight line relative to a second co-ordinate system K', provided that the latter is executing a uniform translatory motion with respect to K. In accordance with the discussion contained in the preceding section, it follows that:

If K is a Galileian co-ordinate system, then every other co-ordinate system K' is a Galileian one, when, in relation to K, it is in a condition of uniform motion of translation. Relative to K' the mechanical laws of Galilei-Newton hold good exactly as they do with respect to K.

We advance a step farther in our generalisation when we express the tenet thus: If, relative to K, K' is a uniformly moving co-ordinate system devoid of rotation, then natural phenomena run their course with respect to K' according to exactly the same general laws as with respect to K. This statement is called the *principle of relativity* (in the restricted sense).

As long as one was convinced that all natural phenomena were capable of representation with the help of classical mechanics, there was no need to doubt the validity of this principle of relativity. But in view of the more recent development of electrodynamics and optics it became more and more evident that classical mechanics affords an insufficient foundation for the physical description of all natural phenomena. At this juncture the question of the validity of the principle of relativity became ripe for discussion, and it did not appear impossible that the answer to this question might be in the negative.

Nevertheless, there are two general facts which at the outset speak very much in favour of the validity of the principle of relativity. Even though classical mechanics does not supply us with a sufficiently broad basis for the theoretical presentation of all physical phenomena, still we must grant it a considerable measure of "truth," since it supplies us with the actual motions of the heavenly bodies with a delicacy of detail little short of wonderful. The principle of relativity must therefore apply with great accuracy in the domain of *mechanics*. But that a principle of such broad generality should hold with such exactness in one domain of phenomena, and yet should be invalid for another, is *a priori* not very probable.

We now proceed to the second argument, to which, moreover, we shall return later. If the principle of relativity (in the restricted sense) does not hold, then the Galileian co-ordinate systems K, K', K'', etc., which are moving uniformly relative to each other, will not be *equivalent* for the description of natural phenomena. In this case we should be constrained to believe that natural laws are capable of being formulated in a particularly simple manner, and of course only on condition that, from amongst all possible Galileian co-ordinate systems, we should have chosen *one* (K_0) of a particular state of motion as our body of reference. We should then be justified (because of its merits for the description of natural phenomena) in calling this system "absolutely at rest," and all other Galileian systems K "in motion." If, for instance, our embankment were the system K_0, then our railway carriage would be a system K, relative to which less simple laws would hold than with respect to K_0. This diminished simplicity would be due to the fact that the carriage K would be in motion (*i.e.* "really") with respect to K_0. In the general laws of nature which have been formulated with reference to K, the magnitude and direction of the velocity of the carriage would necessarily play a part. We should expect, for instance, that the note emitted by an organ-pipe placed with its axis parallel to the direction of travel would be different from that emitted if the axis of the pipe were placed perpendicular to this direction.

Now in virtue of its motion in an orbit round the sun, our earth is comparable with a railway carriage travelling with a velocity of about 30 kilometres per second. If the principle of relativity were not valid we should therefore expect that the direction of motion of the earth at any moment would enter into the laws of nature, and also that physical systems in their behaviour would be dependent on the orientation in space with respect to the earth. For owing to the alteration in direction of the velocity of revolution of the earth in the course of a year, the earth cannot be at rest relative to the hypothetical system K_0 throughout the whole year. However, the most careful observations have never revealed such anisotropic properties in terrestrial physical space, *i.e.* a physical non-equivalence of different directions. This is very powerful argument in favour of the principle of relativity.

VI

THE THEOREM OF THE ADDITION OF VELOCITIES EMPLOYED IN CLASSICAL MECHANICS

LET us suppose our old friend the railway carriage to be travelling along the rails with a constant velocity v, and that a man traverses the length of the carriage in the direction of travel with a velocity w. How quickly or, in other words, with what velocity W does the man advance relative to the embankment during the process? The only possible answer seems to result from the following consideration: If the man were to stand still for a second, he would advance relative to the embankment through a distance v equal numerically to the velocity of the carriage. As a consequence of his walking, however, he traverses an additional distance w relative to the carriage, and hence also relative to the embankment, in this second, the distance w being numerically equal to the velocity with which he is walking. Thus in total he covers the distance $W=v+w$ relative to the embankment in the second considered. We shall see later that this result, which expresses the theorem of the addition of velocities employed in classical mechanics, cannot be maintained; in other words, the law that we have just written down does not hold in reality. For the time being, however, we shall assume its correctness.

VII

THE APPARENT INCOMPATIBILITY OF THE LAW OF PROPAGATION OF LIGHT WITH THE PRINCIPLE OF RELATIVITY

THERE is hardly a simpler law in physics than that according to which light is propagated in empty space. Every child at school knows, or believes he knows, that this propagation takes place in straight lines with a velocity $c = 300,000$ km./sec. At all events we know with great exactness that this velocity is the same for all colours, because if this were not the case, the minimum of emission would not be observed simultaneously for different colours during the eclipse of a fixed star by its dark neighbour. By means of similar considerations based on observations of double stars, the Dutch astronomer De Sitter was also able to show that the velocity of propagation of light cannot depend on the velocity of motion of the body emitting the light. The assumption that this velocity of propagation is dependent on the direction "in space" is in itself improbable.

In short, let us assume that the simple law of the constancy of the velocity of light c (in vacuum) is justifiably believed by the child at school. Who would imagine that this simple law has plunged the conscientiously thoughtful physicist into the greatest intellectual difficulties? Let us consider how these difficulties arise.

Of course we must refer the process of the propagation of light (and indeed every other process) to a rigid reference-body (co-ordinate system). As such a system let us again choose our embankment. We shall imagine the air above it to have been removed. If a ray of light be sent along the embankment, we see from the above that the tip of the ray will be transmitted with the velocity c relative to the embankment. Now let us suppose that our railway carriage is again travelling along the railway lines with the velocity v, and that its direction is the same as that of the ray of light, but its velocity of course much less. Let us inquire about the velocity of propagation of the ray of light relative to the carriage. It is obvious that we can here apply the consideration of the previous section, since the ray of light plays the part of the man walking along relatively to the carriage. The velocity W of the man relative to the embankment is here replaced by the velocity of light relative to the embankment. w is the required velocity of light with respect to the carriage, and we have

$$w = c - v.$$

The velocity of propagation of a ray of light relative to the carriage thus comes out smaller than c.

But this result comes into conflict with the principle of relativity set forth in Section V. For, like every other general law of nature, the law of the transmission of light *in vacuo* must, according to the principle of relativity, be the same for the railway carriage as reference-body as when the rails are the body of reference. But, from our above consideration, this would appear to be impossible. If every ray of light is propagated relative to the embankment with the velocity c, then for this reason it would appear that another law of propagation of light must necessarily hold with respect to the carriage—a result contradictory to the principle of relativity.

In view of this dilemma there appears to be nothing else for it than to abandon either the principle of relativity or the simple law of the propagation of light *in vacuo*. Those of you who have carefully followed the preceding discussion are almost sure to expect that we should retain the principle of relativity, which appeals so convincingly to the intellect because it is so natural and simple. The law of the propagation of light *in vacuo* would then have to be replaced by a more complicated law conformable to the principle of relativity. The development of theoretical physics shows, however, that we cannot pursue this course. The epoch-making theoretical investigations of H. A. Lorentz on the electrodynamical and optical phenomena connected with moving bodies show that experience in this domain leads conclusively to a theory of electromagnetic phenomena, of which the law of the constancy of the velocity of light *in vacuo* is a necessary consequence. Prominent theoretical physicists were therefore more inclined to reject the principle of relativity, in spite of the fact that no empirical data had been found which were contradictory to this principle.

At this juncture the theory of relativity entered the arena. As a result of an analysis of the physical conceptions of time and space, it became evident that *in reality there is not the least incompatibility between the principle of relativity and the law of propagation of light,* and that by systematically holding fast to both these laws a logically rigid theory could be arrived at. This theory has been called the *special theory of relativity* to distinguish it from the extended theory, with which we shall deal later. In the following pages we shall present the fundamental ideas of the special theory of relativity.

VIII

ON THE IDEA OF TIME IN PHYSICS

LIGHTNING has struck the rails on our railway embankment at two places *A* and *B* far distant from each other. I make the additional assertion that these two lightning flashes occurred simultaneously. If I ask you whether there is sense in this statement, you will answer my question with a decided "Yes." But if I now approach you with the request to explain to me the sense of the statement more precisely, you find after some consideration that the answer to this question is not so easy as it appears at first sight.

After some time perhaps the following answer would occur to you: "The significance of the statement is clear in itself and needs no further explanation; of course it would require some consideration if I were to be commissioned to determine by observations whether in the actual case the two events took place simultaneously or not." I cannot be satisfied with this answer for the following reason. Supposing that as a result of ingenious considerations an able meteorologist were to discover that the lightning must always strike the places *A* and *B* simultaneously, then we should be faced with the task of testing whether or not this theoretical result is in accordance with the reality. We encounter the same difficulty with all physical statements in which the conception "simultaneous" plays a part. The concept does not exist for the physicist until he has the possibility of discovering whether or not it is fulfilled in an actual case. We thus require a definition of simultaneity such that this definition supplies us with the method by means of which, in the present case, he can decide by experiment whether or not both the lightning strokes occurred simultaneously. As long as this requirement is not satisfied, I allow myself to be deceived as a physicist (and of course the same applies if I am not a physicist), when I imagine that I am able to attach a meaning to the statement of simultaneity. (I would ask the reader not to proceed farther until he is fully convinced on this point.)

After thinking the matter over for some time you then offer the following suggestion with which to test simultaneity. By measuring along the rails, the connecting line AB should be measured up and an observer placed at the mid-point M of the distance AB. This observer should be supplied with an arrangement (*e.g.* two mirrors inclined at 90°) which allows him visually to observe both places A and B at the same time. If the observer perceives the two flashes of lightning at the same time, then they are simultaneous.

I am very pleased with this suggestion, but for all that I cannot regard the matter as quite settled, because I feel constrained to raise the following objection: "Your definition would certainly be right, if I only knew that the light by means of which the observer at M perceives the lightning flashes travels along the length $A \longrightarrow M$ with the same velocity as along the length $B \longrightarrow M$. But an examination of this suppositition would only be possible if we already had at our disposal the means of measuring time. It would thus appear as though we were moving here in a logical circle."

After further consideration you cast a somewhat disdainful glance at me—and rightly so—and you declare: "I maintain my previous definition nevertheless, because in reality it assumes absolutely nothing about light. There is only *one* demand to be made of the definition of simultaneity, namely, that in every real case it must supply us with an empirical decision as to whether or not the conception that has to be defined is fulfilled. That my definition satisfies this demand is indisputable. That light requires the same time to traverse the path $A \longrightarrow M$ as for the path $B \longrightarrow M$ is in reality neither a *supposition nor a hypothesis* about the physical nature of light, but a *stipulation* which I can make of my own freewill in order to arrive at a definition of simultaneity."

It is clear that this definition can be used to give an exact meaning not only to *two* events, but to as many events as we care to choose, and independently of the positions of the scenes of the events with respect to the body of reference[1] (here the railway embankment). We are thus led also to a definition of "time" in physics. For this purpose we suppose that clocks of identical construction are placed at the points A, B and C of

[1] We suppose further, that, when three events A, B and C occur in different places in such a manner that A is simultaneous with B, and B is simultaneous with C (simultaneous in the sense of the above definition), then the criterion for the simultaneity of the pair of events A, C is also satisfied. This assumption is a physical hypothesis about the law of propagation of light; it must certainly be fulfilled if we are to maintain the law of the constancy of the velocity of light *in vacuo*.

the railway line (co-ordinate system), and that they are set in such a manner that the positions of their pointers are simultaneously (in the above sense) the same. Under these conditions we understand by the "time" of an event the reading (position of the hands) of that one of these clocks which is in the immediate vicinity (in space) of the event. In this manner a time-value is associated with every event which is essentially capable of observation.

This stipulation contains a further physical hypothesis, the validity of which will hardly be doubted without empirical evidence to the contrary. It has been assumed that all these clocks go *at the same rate* if they are of identical construction. Stated more exactly: When two clocks arranged at rest in different places of a reference-body are set in such a manner that a *particular* position of the pointers of the one clock is *simultaneous* (in the above sense) with the *same* position of the pointers of the other clock, then identical "settings" are always simultaneous (in the sense of the above definition).

IX

THE RELATIVITY OF SIMULTANEITY

UP to now our considerations have been referred to a particular body of reference, which we have styled a "railway embankment." We suppose a very long train travelling along the rails with the constant velocity v and in the direction indicated in Fig. 1. People travelling in this train will with advantage use the train as a rigid reference-body (co-ordinate system); they regard all events in

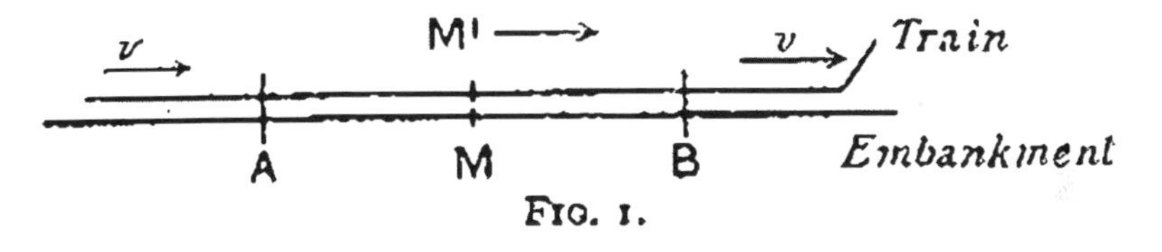

FIG. 1.

reference to the train. Then every event which takes place along the line also takes place at a particular point of the train. Also the definition of simultaneity can be given relative to the train in exactly the same way as with respect to the embankment. As a natural consequence, however, the following question arises :

Are two events (*e.g.* the two strokes of lightning A and B) which are simultaneous *with reference to the railway embankment* also simultaneous *relatively to the train* ? We shall show directly that the answer must be in the negative.

When we say that the lightning strokes A and B are

simultaneous with respect to the embankment, we mean: the rays of light emitted at the places A and B, where the lightning occurs, meet each other at the mid-point M of the length $A \longrightarrow B$ of the embankment. But the events A and B also correspond to positions A and B on the train. Let M' be the mid-point of the distance $A \longrightarrow B$ on the travelling train. Just when the flashes[1] of lightning occur, this point M' naturally coincides with the point M, but it moves towards the right in the diagram with the velocity v of the train. If an observer sitting in the position M' in the train did not possess this velocity, then he would remain permanently at M, and the light rays emitted by the flashes of lightning A and B would reach him simultaneously, *i.e.* they would meet just where he is situated. Now in reality (considered with reference to the railway embankment) he is hastening towards the beam of light coming from B, whilst he is riding on ahead of the beam of light coming from A. Hence the observer will see the beam of light emitted from B earlier than he will see that emitted from A. Observers who take the railway train as their reference-body must therefore come to the conclusion that the lightning flash B took place earlier than the lightning flash A. We thus arrive at the important result:

Events which are simultaneous with reference to the embankment are not simultaneous with respect to the train, and *vice versa* (relativity of simultaneity). Every reference-body (co-ordinate system) has its own particular time; unless we are told the reference-body to which the statement of time refers, there is no meaning in a statement of the time of an event.

[1] As judged from the embankment.

Now before the advent of the theory of relativity it had always tacitly been assumed in physics that the statement of time had an absolute significance, *i.e.* that it is independent of the state of motion of the body of reference. But we have just seen that this assumption is incompatible with the most natural definition of simultaneity; if we discard this assumption, then the conflict between the law of the propagation of light *in vacuo* and the principle of relativity (developed in Section VII) disappears.

We were led to that conflict by the considerations of Section VI, which are now no longer tenable. In that section we concluded that the man in the carriage, who traverses the distance *w per second* relative to the carriage, traverses the same distance also with respect to the embankment *in each second* of time. But, according to the foregoing considerations, the time required by a particular occurrence with respect to the carriage must not be considered equal to the duration of the same occurrence as judged from the embankment (as reference-body). Hence it cannot be contended that the man in walking travels the distance *w* relative to the railway line in a time which is equal to one second as judged from the embankment.

Moreover, the considerations of Section VI are based on yet a second assumption, which, in the light of a strict consideration, appears to be arbitrary, although it was always tacitly made even before the introduction of the theory of relativity.

X

ON THE RELATIVITY OF THE CONCEPTION OF DISTANCE

LET us consider two particular points on the train[1] travelling along the embankment with the velocity v, and inquire as to their distance apart. We already know that it is necessary to have a body of reference for the measurement of a distance, with respect to which body the distance can be measured up. It is the simplest plan to use the train itself as reference-body (co-ordinate system). An observer in the train measures the interval by marking off his measuring-rod in a straight line (*e.g.* along the floor of the carriage) as many times as is necessary to take him from the one marked point to the other. Then the number which tells us how often the rod has to be laid down is the required distance.

It is a different matter when the distance has to be judged from the railway line. Here the following method suggests itself. If we call A' and B' the two points on the train whose distance apart is required, then both of these points are moving with the velocity v along the embankment. In the first place we require to determine the points A and B of the embankment which are just being passed by the two points A' and B' at a

[1] *e.g.* the middle of the first and of the hundredth carriage.

particular time t—judged from the embankment. These points A and B of the embankment can be determined by applying the definition of time given in Section VIII. The distance between these points A and B is then measured by repeated application of the measuring-rod along the embankment.

A priori it is by no means certain that this last measurement will supply us with the same result as the first. Thus the length of the train as measured from the embankment may be different from that obtained by measuring in the train itself. This circumstance leads us to a second objection which must be raised against the apparently obvious consideration of Section VI. Namely, if the man in the carriage covers the distance w in a unit of time—*measured from the train,*—then this distance—*as measured from the embankment*—is not necessarily also equal to w.

XI

THE LORENTZ TRANSFORMATION

THE results of the last three sections show that the apparent incompatibility of the law of propagation of light with the principle of relativity (Section VII) has been derived by means of a consideration which borrowed two unjustifiable hypotheses from classical mechanics; these are as follows:

(1) The time-interval (time) between two events is independent of the condition of motion of the body of reference.

(2) The space-interval (distance) between two points of a rigid body is independent of the condition of motion of the body of reference.

If we drop these hypotheses, then the dilemma of Section VII disappears, because the theorem of the addition of velocities derived in Section VI becomes invalid. The possibility presents itself that the law of the propagation of light *in vacuo* may be compatible with the principle of relativity, and the question arises: How have we to modify the considerations of Section VI in order to remove the apparent disagreement between these two fundamental results of experience? This question leads to a general one. In the discussion of

Section VI we have to do with places and times relative both to the train and to the embankment. How are we to find the place and time of an event in relation to the train, when we know the place and time of the event with respect to the railway embankment? Is there a thinkable answer to this question of such a nature that the law of transmission of light *in vacuo* does not contradict the principle of relativity? In other words: Can we conceive of a relation between place and time of the individual events relative to both reference-bodies, such that every ray of light possesses the velocity of transmission c relative to the embankment and relative to the train? This question leads to a quite definite positive answer, and to a perfectly definite transformation law for the space-time magnitudes of an event when changing over from one body of reference to another.

Before we deal with this, we shall introduce the following incidental consideration. Up to the present we have only considered events taking place along the embankment, which had mathematically to assume the function of a straight line. In the manner indicated in Section II we can imagine this reference-body supplemented laterally and in a vertical direction by means of a framework of rods, so that an event which takes place anywhere can be localised with reference to this framework. Similarly, we can imagine the train travelling with the velocity v to be continued across the whole of space, so that every event, no matter how far off it may be, could also be localised with respect to the second framework. Without committing any fundamental error, we can disregard the fact that in reality these frameworks would continually interfere with each other, owing to the impenetrability of solid bodies. In every such framework we imagine three surfaces perpendicular to each other marked out, and designated as "co-ordinate planes" ("co-ordinate system"). A co-ordinate system K then corresponds to the embankment, and a co-ordinate system K' to the train. An event, wherever it may have taken place, would be fixed in space with respect to K by the three perpendiculars x, y, z on the co-ordinate planes, and with regard to time by a time-value t. Relative to K', *the same event* would be fixed in respect of space and time by corresponding values x', y', z', t', which of course are not identical with x, y, z, t. It has already been set forth in detail how these magnitudes are to be regarded as results of physical measurements.

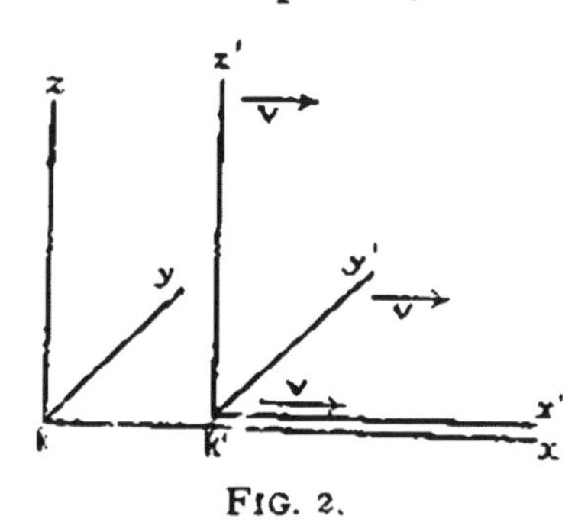

FIG. 2.

Obviously our problem can be exactly formulated in the following manner. What are the values x', y', z', t', of an event with respect to K', when the magnitudes x, y, z, t, of the same event with respect to K are given? The relations must be so chosen that the law of the transmission of light *in vacuo* is satisfied for one and the same ray of light (and of course for every ray) with respect to K and K'. For the relative orientation in space of the co-ordinate systems indicated in the diagram (Fig. 2), this problem is solved by means of the equations:

$$x' = \frac{x - vt}{\sqrt{1 - \frac{v^2}{c^2}}}$$

$$y' = y$$
$$z' = z$$
$$t' = \frac{t - \frac{v}{c^2} \cdot x}{\sqrt{1 - \frac{v^2}{c^2}}}$$

This system of equations is known as the "Lorentz transformation."[1]

If in place of the law of transmission of light we had taken as our basis the tacit assumptions of the older mechanics as to the absolute character of times and lengths, then instead of the above we should have obtained the following equations:

$$x' = x - vt$$
$$y' = y$$
$$z' = z$$
$$t' = t.$$

This system of equations is often termed the "Galilei transformation." The Galilei transformation can be obtained from the Lorentz transformation by substituting an infinitely large value for the velocity of light c in the latter transformation.

Aided by the following illustration, we can readily see that, in accordance with the Lorentz transformation, the law of the transmission of light *in vacuo* is satisfied both for the reference-body K and for the reference-body K'. A light-signal is sent along the positive x-axis, and this light-stimulus advances in accordance with the equation

$$x = ct,$$

[1] A simple derivation of the Lorentz transformation is given in Appendix I.

i.e. with the velocity c. According to the equations of the Lorentz transformation, this simple relation between x and t involves a relation between x' and t'. In point of fact, if we substitute for x the value ct in the first and fourth equations of the Lorentz transformation, we obtain :

$$x' = \frac{(c - v)t}{\sqrt{1 - \frac{v^2}{c^2}}}$$

$$t' = \frac{\left(1 - \frac{v}{c}\right)t}{\sqrt{1 - \frac{v^2}{c^2}}},$$

from which, by division, the expression

$$x' = ct'$$

immediately follows. If referred to the system K', the propagation of light takes place according to this equation. We thus see that the velocity of transmission relative to the reference-body K' is also equal to c. The same result is obtained for rays of light advancing in any other direction whatsoever. Of course this is not surprising, since the equations of the Lorentz transformation were derived conformably to this point of view.

XII

THE BEHAVIOUR OF MEASURING-RODS AND CLOCKS IN MOTION

I PLACE a metre-rod in the x'-axis of K' in such a manner that one end (the beginning) coincides with the point $x'=0$, whilst the other end (the end of the rod) coincides with the point $x'=1$. What is the length of the metre-rod relatively to the system K? In order to learn this, we need only ask where the beginning of the rod and the end of the rod lie with respect to K at a particular time t of the system K. By means of the first equation of the Lorentz transformation the values of these two points at the time $t=0$ can be shown to be

$$x_{(\text{beginning of rod})} = 0 . \sqrt{1-\frac{v^2}{c^2}}$$

$$x_{(\text{end of rod})} = 1 . \sqrt{1-\frac{v^2}{c^2}},$$

the distance between the points being $\sqrt{1-\frac{v^2}{c^2}}$. But the metre-rod is moving with the velocity v relative to K. It therefore follows that the length of a rigid metre-rod moving in the direction of its length with a velocity v is $\sqrt{1-v^2/c^2}$ of a metre. The rigid rod is thus shorter when in motion than when at rest, and the more quickly it is moving, the shorter is the rod. For the velocity $v=c$ we should have $\sqrt{1-v^2/c^2}=0$, and for still greater velocities the square-root becomes imaginary. From this we conclude that in the theory of relativity the velocity c plays the part of a limiting velocity, which can neither be reached nor exceeded by any real body.

Of course this feature of the velocity c as a limiting velocity also clearly follows from the equations of the Lorentz transformation, for these become meaningless if we choose values of v greater than c.

If, on the contrary, we had considered a metre-rod at rest in the x-axis with respect to K, then we should have found that the length of the rod as judged from K' would have been $\sqrt{1-v^2/c^2}$; this is quite in accordance with the principle of relativity which forms the basis of our considerations.

A priori it is quite clear that we must be able to learn something about the physical behaviour of measuring-rods and clocks from the equations of transformation, for the magnitudes x, y, z, t, are nothing more nor less than the results of measurements obtainable by means of measuring-rods and clocks. If we had based our considerations on the Galilei transformation we should not have obtained a contraction of the rod as a consequence of its motion.

Let us now consider a seconds-clock which is permanently situated at the origin ($x'=0$) of K'. $t'=0$ and $t'=1$ are two successive ticks of this clock. The first and fourth equations of the Lorentz transformation give for these two ticks:

$$t=0$$

and

$$t=\frac{1}{\sqrt{1-\frac{v^2}{c^2}}}.$$

As judged from K, the clock is moving with the velocity v; as judged from this reference-body, the time which elapses between two strokes of the clock is not one second, but $\frac{1}{\sqrt{1-\frac{v^2}{c^2}}}$ seconds, *i.e.* a somewhat larger time. As a consequence of its motion the clock goes more slowly than when at rest. Here also the velocity c plays the part of an unattainable limiting velocity.

XIII

THEOREM OF THE ADDITION OF VELOCITIES. THE EXPERIMENT OF FIZEAU

NOW in practice we can move clocks and measuring-rods only with velocities that are small compared with the velocity of light; hence we shall hardly be able to compare the results of the previous section directly with the reality. But, on the other hand, these results must strike you as being very singular, and for that reason I shall now draw another conclusion from the theory, one which can easily be derived from the foregoing considerations, and which has been most elegantly confirmed by experiment.

In Section VI we derived the theorem of the addition of velocities in one direction in the form which also results from the hypotheses of classical mechanics. This theorem can also be deduced readily from the Galilei transformation (Section XI). In place of the man walking inside the carriage, we introduce a point moving relatively to the co-ordinate system K' in accordance with the equation

$$x' = wt'.$$

By means of the first and fourth equations of the Galilei transformation we can express x' and t' in terms of x and t, and we then obtain

$$x = (v + w)t.$$

This equation expresses nothing else than the law of motion of the point with reference to the system K (of the man with reference to the embankment). We denote this velocity by the symbol W, and we then obtain, as in Section VI,

$$W = v + w \quad . \quad . \quad . \quad (A).$$

But we can carry out this consideration just as well on the basis of the theory of relativity. In the equation

$$x' = wt'$$

we must then express x' and t' in terms of x and t, making use of the first and fourth equations of the *Lorentz transformation*. Instead of the equation (A) we then obtain the equation

$$W = \frac{v + w}{1 + \frac{vw}{c^2}} \quad . \quad . \quad . \quad (B),$$

which corresponds to the theorem of addition for velocities in one direction according to the theory of relativity. The question now arises as to which of these two theorems is the better in accord with experience. On this point we are enlightened by a most important experiment which the brilliant physicist Fizeau performed more than half a century ago, and which has been repeated since then by some of the best experimental physicists, so that there can be no doubt about its result. The experiment is concerned with the following question. Light travels in a motionless liquid with a particular velocity w. How quickly does it travel in the direction of the arrow in the tube T (see the accompanying diagram, Fig. 3) when the liquid above mentioned is flowing through the tube with a velocity v?

In accordance with the principle of relativity we shall certainly have to take for granted that the propagation of light always takes place with the same velocity w *with respect to the liquid*, whether the latter is in motion with reference to other bodies or not. The velocity of light relative to the liquid and the velocity of the latter relative to the tube are thus known, and we require the velocity of light relative to the tube.

It is clear that we have the problem of Section VI again before us. The tube plays the part of the railway embankment or of the co-ordinate system K, the liquid plays the part of the carriage or of the co-ordinate system K', and finally, the light plays the part of the

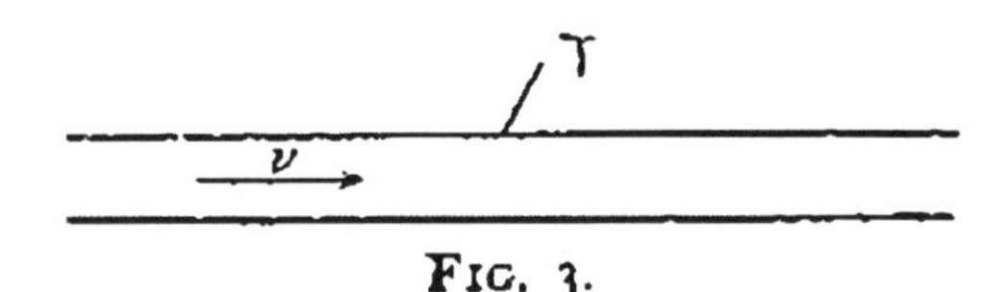

Fig. 3.

man walking along the carriage, or of the moving point in the present section. If we denote the velocity of the light relative to the tube by W, then this is given by the equation (A) or (B), according as the Galilei transformation or the Lorentz transformation corresponds to the facts. Experiment[1] decides in favour of equation (B) derived from the theory of relativity, and the agreement is, indeed, very exact. According to

[1] Fizeau found $W = w + v\left(1 - \frac{1}{n^2}\right)$, where $n = \frac{c}{w}$ is the index of refraction of the liquid. On the other hand, owing to the smallness of $\frac{vw}{c^2}$ as compared with 1, we can replace (B) in the first place by $W = (w + v)\left(1 - \frac{vw}{c^2}\right)$, or to the same order of approximation by $w + v\left(1 - \frac{1}{n^2}\right)$, which agrees with Fizeau's result.

recent and most excellent measurements by Zeeman, the influence of the velocity of flow v on the propagation of light is represented by formula (B) to within one per cent.

Nevertheless we must now draw attention to the fact that a theory of this phenomenon was given by H. A. Lorentz long before the statement of the theory of relativity. This theory was of a purely electrodynamical nature, and was obtained by the use of particular hypotheses as to the electromagnetic structure of matter. This circumstance, however, does not in the least diminish the conclusiveness of the experiment as a crucial test in favour of the theory of relativity, for the electrodynamics of Maxwell-Lorentz, on which the original theory was based, in no way opposes the theory of relativity. Rather has the latter been developed from electrodynamics as an astoundingly simple combination and generalisation of the hypotheses, formerly independent of each other, on which electrodynamics was built.

XIV

THE HEURISTIC VALUE OF THE THEORY OF RELATIVITY

OUR train of thought in the foregoing pages can be epitomised in the following manner. Experience has led to the conviction that, on the one hand, the principle of relativity holds true, and that on the other hand the velocity of transmission of light *in vacuo* has to be considered equal to a constant c. By uniting these two postulates we obtained the law of transformation for the rectangular co-ordinates x, y, z and the time t of the events which constitute the processes of nature. In this connection we did not obtain the Galilei transformation, but, differing from classical mechanics, the *Lorentz transformation.*

The law of transmission of light, the acceptance of which is justified by our actual knowledge, played an important part in this process of thought. Once in possession of the Lorentz transformation, however, we can combine this with the principle of relativity, and sum up the theory thus :

Every general law of nature must be so constituted that it is transformed into a law of exactly the same form when, instead of the space-time variables x, y, z, t of the original co-ordinate system K, we introduce new space-time variables x', y', z', t' of a co-ordinate system

K'. In this connection the relation between the ordinary and the accented magnitudes is given by the Lorentz transformation. Or, in brief: General laws of nature are co-variant with respect to Lorentz transformations.

This is a definite mathematical condition that the theory of relativity demands of a natural law, and in virtue of this, the theory becomes a valuable heuristic aid in the search for general laws of nature. If a general law of nature were to be found which did not satisfy this condition, then at least one of the two fundamental assumptions of the theory would have been disproved. Let us now examine what general results the latter theory has hitherto evinced.

XV

GENERAL RESULTS OF THE THEORY

IT is clear from our previous considerations that the (special) theory of relativity has grown out of electrodynamics and optics. In these fields it has not appreciably altered the predictions of theory, but it has considerably simplified the theoretical structure, *i.e.* the derivation of laws, and—what is incomparably more important—it has considerably reduced the number of independent hypotheses forming the basis of theory. The special theory of relativity has rendered the Maxwell-Lorentz theory so plausible, that the latter would have been generally accepted by physicists even if experiment had decided less unequivocally in its favour.

Classical mechanics required to be modified before it could come into line with the demands of the special theory of relativity. For the main part, however, this modification affects only the laws for rapid motions, in which the velocities of matter v are not very small as compared with the velocity of light. We have experience of such rapid motions only in the case of electrons and ions; for other motions the variations from the laws of classical mechanics are too small to make themselves evident in practice. We shall not consider the motion of stars until we come to speak of the general theory of relativity. In accordance with the theory of relativity he kinetic energy of a material point of mass m is no longer given by the well-known expression

$$m\frac{v^2}{2},$$

ut by the expression

$$\frac{mc^2}{\sqrt{1-\frac{v^2}{c^2}}}.$$

This expression approaches infinity as the velocity v pproaches the velocity of light c. The velocity must herefore always remain less than c, however great may e the energies used to produce the acceleration. If /e develop the expression for the kinetic energy in the orm of a series, we obtain

$$mc^2+m\frac{v^2}{2}+\frac{3}{8}m\frac{v^4}{c^2}+\ldots.$$

When $\frac{v^2}{c^2}$ is small compared with unity, the third f these terms is always small in comparison with the econd, which last is alone considered in classical nechanics. The first term mc^2 does not contain he velocity, and requires no consideration if we are only lealing with the question as to how the energy of a oint-mass depends on the velocity. We shall speak f its essential significance later.

The most important result of a general character to which the special theory of relativity has led is concerned with the conception of mass. Before the advent of relativity, physics recognised two conservation laws of undamental importance, namely, the law of the conservation of energy and the law of the conservation of mass; these two fundamental laws appeared to be quite independent of each other. By means of the theory of relativity they have been united into one law. We shall now briefly consider how this unification came about, and what meaning is to be attached to it.

The principle of relativity requires that the law of the conservation of energy should hold not only with reference to a co-ordinate system K, but also with respect to every co-ordinate system K' which is in a state of uniform motion of translation relative to K, or, briefly, relative to every "Galileian" system of co-ordinates. In contrast to classical mechanics, the Lorentz transformation is the deciding factor in the transition from one such system to another.

By means of comparatively simple considerations we are led to draw the following conclusion from these premises, in conjunction with the fundamental equations of the electrodynamics of Maxwell: A body moving with the velocity v, which absorbs[1] an amount of energy E_0 in the form of radiation without suffering an alteration in velocity in the process, has, as a consequence, its energy increased by an amount

$$\frac{E_0}{\sqrt{1-\frac{v^2}{c^2}}}.$$

In consideration of the expression given above for the kinetic energy of the body, the required energy of the body comes out to be

$$\frac{\left(m+\frac{E_0}{c^2}\right)c^2}{\sqrt{1-\frac{v^2}{c^2}}}.$$

[1] E_0 is the energy taken up, as judged from a co-ordinate system moving with the body.

Thus the body has the same energy as a body of mass $\left(m+\frac{E_0}{c^2}\right)$ moving with the velocity v. Hence we can say: If a body takes up an amount of energy E_0, then its inertial mass increases by an amount $\frac{E_0}{c^2}$; the inertial mass of a body is not a constant, but varies according to the change in the energy of the body. The inertial mass of a system of bodies can even be regarded as a measure of its energy. The law of the conservation of the mass of a system becomes identical with the law of the conservation of energy, and is only valid provided that the system neither takes up nor sends out energy. Writing the expression for the energy in the form

$$\frac{mc^2+E_0}{\sqrt{1-\frac{v^2}{c^2}}},$$

we see that the term mc^2, which has hitherto attracted our attention, is nothing else than the energy possessed by the body [1] before it absorbed the energy E_0.

A direct comparison of this relation with experiment is not possible at the present time, owing to the fact that the changes in energy E_0 to which we can subject a system are not large enough to make themselves perceptible as a change in the inertial mass of the system. $\frac{E_0}{c^2}$ is too small in comparison with the mass m, which was present before the alteration of the energy. It is owing to this circumstance that classical mechanics was able to establish successfully the conservation of mass as a law of independent validity.

[1] As judged from a co-ordinate system moving with the body.

Let me add a final remark of a fundamental nature. The success of the Faraday-Maxwell interpretation of electromagnetic action at a distance resulted in physicists becoming convinced that there are no such things as instantaneous actions at a distance (not involving an intermediary medium) of the type of Newton's law of gravitation. According to the theory of relativity, action at a distance with the velocity of light always takes the place of instantaneous action at a distance or of action at a distance with an infinite velocity of transmission. This is connected with the fact that the velocity c plays a fundamental rôle in this theory. In Part II we shall see in what way this result becomes modified in the general theory of relativity.

XVI

EXPERIENCE AND THE SPECIAL THEORY OF RELATIVITY

TO what extent is the special theory of relativity supported by experience? This question is not easily answered for the reason already mentioned in connection with the fundamental experiment of Fizeau. The special theory of relativity has crystallised out from the Maxwell-Lorentz theory of electromagnetic phenomena. Thus all facts of experience which support the electromagnetic theory also support the theory of relativity. As being of particular importance, I mention here the fact that the theory of relativity enables us to predict the effects produced on the light reaching us from the fixed stars. These results are obtained in an exceedingly simple manner, and the effects indicated, which are due to the relative motion of the earth with reference to those fixed stars, are found to be in accord with experience. We refer to the yearly movement of the apparent position of the fixed stars resulting from the motion of the earth round the sun (aberration), and to the influence of the radial components of the relative motions of the fixed stars with respect to the earth on the colour of the light reaching us from them. The latter effect manifests itself in a slight displacement of the spectral lines of the light transmitted to us from a fixed star, as compared with the position of the same spectral lines when they are produced by a terrestrial source of light (Doppler principle). The experimental arguments in favour of the Maxwell-Lorentz theory, which are at the same time arguments in favour of the theory of relativity, are too numerous to be set forth here. In reality they limit the theoretical possibilities to such an extent, that no other theory than that of Maxwell and Lorentz has been able to hold its own when tested by experience.

But there are two classes of experimental facts hitherto obtained which can be represented in the Maxwell-Lorentz theory only by the introduction of an auxiliary hypothesis, which in itself—*i.e.* without making use of the theory of relativity—appears extraneous.

It is known that cathode rays and the so-called β-rays emitted by radioactive substances consist of negatively electrified particles (electrons) of very small inertia and large velocity. By examining the deflection of these rays under the influence of electric and magnetic fields, we can study the law of motion of these particles very exactly.

In the theoretical treatment of these electrons, we are faced with the difficulty that electrodynamic theory of itself is unable to give an account of their nature. For since electrical masses of one sign repel each other, the negative electrical masses constituting the electron would necessarily be scattered under the influence of their mutual repulsions, unless there are forces of another kind operating between them, the nature of which has hitherto remained obscure to us.[1] If we now assume that the relative distances between the electrical masses constituting the electron remain unchanged during the motion of the electron (rigid connection in the sense of classical mechanics), we arrive at a law of motion of the electron which does not agree with experience. Guided by purely formal points of view, H. A. Lorentz was the first to introduce the hypothesis that the particles constituting the electron experience a contraction in the direction of motion in consequence of that motion, the amount of this contraction being proportional to the expression $\sqrt{1-\frac{v^2}{c^2}}$. This hypothesis, which is not justifiable by any electrodynamical facts, supplies us then with that particular law of motion which has been confirmed with great precision in recent years.

The theory of relativity leads to the same law of motion, without requiring any special hypothesis whatsoever as to the structure and the behaviour of the electron. We arrived at a similar conclusion in Section XIII in connection with the experiment of Fizeau, the result of which is foretold by the theory of relativity without the necessity of drawing on hypotheses as to the physical nature of the liquid.

The second class of facts to which we have alluded has reference to the question whether or not the motion of the earth in space can be made perceptible in terrestrial experiments. We have already remarked in Section V that all attempts of this nature led to a negative result. Before the theory of relativity was put forward, it was

[1] The general theory of relativity renders it likely that the electrical masses of an electron are held together by gravitational forces.

difficult to become reconciled to this negative result, for reasons now to be discussed. The inherited prejudices about time and space did not allow any doubt to arise as to the prime importance of the Galilei transformation for changing over from one body of reference to another. Now assuming that the Maxwell-Lorentz equations hold for a reference-body K, we then find that they do not hold for a reference-body K' moving uniformly with respect to K, if we assume that the relations of the Galileian transformation exist between the co-ordinates of K and K'. It thus appears that of all Galileian co-ordinate systems one (K) corresponding to a particular state of motion is physically unique. This result was interpreted physically by regarding K as at rest with respect to a hypothetical æther of space. On the other hand, all co-ordinate systems K' moving relatively to K were to be regarded as in motion with respect to the æther. To this motion of K' against the æther ("æther-drift" relative to K') were assigned the more complicated laws which were supposed to hold relative to K'. Strictly speaking, such an æther-drift ought also to be assumed relative to the earth, and for a long time the efforts of physicists were devoted to attempts to detect the existence of an æther-drift at the earth's surface.

In one of the most notable of these attempts Michelson devised a method which appears as though it must be decisive. Imagine two mirrors so arranged on a rigid body that the reflecting surfaces face each other. A ray of light requires a perfectly definite time T to pass from one mirror to the other and back again, if the whole system be at rest with respect to the æther. It is found by calculation, however, that a slightly different time

T' is required for this process, if the body, together with the mirrors, be moving relatively to the æther. And yet another point: it is shown by calculation that for a given velocity v with reference to the æther, this time T' is different when the body is moving perpendicularly to the planes of the mirrors from that resulting when the motion is parallel to these planes. Although the estimated difference between these two times is exceedingly small, Michelson and Morley performed an experiment involving interference in which this difference should have been clearly detectable. But the experiment gave a negative result—a fact very perplexing to physicists. Lorentz and FitzGerald rescued the theory from this difficulty by assuming that the motion of the body relative to the æther produces a contraction of the body in the direction of motion, the amount of contraction being just sufficient to compensate for the difference in time mentioned above. Comparison with the discussion in Section XII shows that also from the standpoint of the theory of relativity this solution of the difficulty was the right one. But on the basis of the theory of relativity the method of interpretation is incomparably more satisfactory. According to this theory there is no such thing as a "specially favoured" (unique) co-ordinate system to occasion the introduction of the æther-idea, and hence there can be no æther-drift, nor any experiment with which to demonstrate it. Here the contraction of moving bodies follows from the two fundamental principles of the theory without the introduction of particular hypotheses; and as the prime factor involved in this contraction we find, not the motion in itself, to which we cannot attach any meaning, but the motion with respect to the body of reference chosen in the particular case in point. Thus for a co-ordinate system moving with the earth the mirror system of Michelson and Morley is not shortened, but it *is* shortened for a co-ordinate system which is at rest relatively to the sun.

XVII

MINKOWSKI'S FOUR-DIMENSIONAL SPACE

THE non-mathematician is seized by a mysterious shuddering when he hears of "four-dimensional" things, by a feeling not unlike that awakened by thoughts of the occult. And yet there is no more common-place statement than that the world in which we live is a four-dimensional space-time continuum.

Space is a three-dimensional continuum. By this we mean that it is possible to describe the position of a point (at rest) by means of three numbers (co-ordinates) x, y, z, and that there is an indefinite number of points in the neighbourhood of this one, the position of which can be described by co-ordinates such as x_1, y_1, z_1, which may be as near as we choose to the respective values of the co-ordinates x, y, z of the first point. In virtue of the latter property we speak of a "continuum," and owing to the fact that there are three co-ordinates we speak of it as being "three-dimensional."

Similarly, the world of physical phenomena which was briefly called "world" by Minkowski is naturally four-dimensional in the space-time sense. For it is composed of individual events, each of which is described by four numbers, namely, three space co-ordinates x, y, z and a time co-ordinate, the time-value t. The "world" is in this sense also a continuum; for to every event there are as many "neighbouring" events (realised or at least thinkable) as we care to choose, the co-ordinates x_1, y_1, z_1, t_1 of which differ by an indefinitely small amount from those of the event x, y, z, t originally considered. That we have not been accustomed to regard the world in this sense as a four-dimensional continuum is due to the fact that in physics, before the advent of the theory of relativity, time played a different and more independent rôle, as compared with the space co-ordinates. It is for this reason that we have been in the habit of treating time as an independent continuum. As a matter of fact, according to classical mechanics, time is absolute, *i.e.* it is independent of the position and the condition of motion of the system of co-ordinates. We see this expressed in the last equation of the Galileian transformation ($t' = t$).

The four-dimensional mode of consideration of the "world" is natural on the theory of relativity, since according to this theory time is robbed of its independence. This is shown by the fourth equation of the Lorentz transformation:

$$t' = \frac{t - \frac{v}{c^2}x}{\sqrt{1 - \frac{v^2}{c^2}}}.$$

Moreover, according to this equation the time difference $\Delta t'$ of two events with respect to K' does not in general vanish, even when the time difference Δt of the same events with reference to K vanishes. Pure "space-distance" of two events with respect to K results in "time-distance" of the same events with respect to K'. But the discovery of Minkowski, which was of importance for the formal development of the theory of relativity, does not lie here. It is to be found rather in the fact of his recognition that the four-dimensional space-time continuum of the theory of relativity, in its most essential formal properties, shows a pronounced relationship to the three-dimensional continuum of Euclidean geometrical space.[1] In order to give due prominence to this relationship, however, we must replace the usual time co-ordinate t by an imaginary magnitude $\sqrt{-1}\,.\,ct$ proportional to it. Under these conditions, the natural laws satisfying the demands of the (special) theory of relativity assume mathematical forms, in which the time co-ordinate plays exactly the same rôle as the three space co-ordinates. Formally, these four co-ordinates correspond exactly to the three space co-ordinates in Euclidean geometry. It must be clear even to the non-mathematician that, as a consequence of this purely formal addition to our knowledge, the theory perforce gained clearness in no mean measure.

These inadequate remarks can give the reader only a vague notion of the important idea contributed by Minkowski. Without it the general theory of relativity, of which the fundamental ideas are developed in the following pages, would perhaps have got no farther than its long clothes. Minkowski's work is doubtless difficult of access to anyone inexperienced in mathematics, but since it is not necessary to have a very exact grasp of this work in order to understand the fundamental ideas of either the special or the general theory of relativity, I shall at present leave it here, and shall revert to it only towards the end of Part II.

[1] Cf. the somewhat more detailed discussion in Appendix II.

PART II

THE GENERAL THEORY OF RELATIVITY

XVIII

SPECIAL AND GENERAL PRINCIPLE OF RELATIVITY

THE basal principle, which was the pivot of all our previous considerations, was the *special* principle of relativity, *i.e.* the principle of the physical relativity of all *uniform* motion. Let us once more analyse its meaning carefully.

It was at all times clear that, from the point of view of the idea it conveys to us, every motion must only be considered as a relative motion. Returning to the illustration we have frequently used of the embankment and the railway carriage, we can express the fact of the motion here taking place in the following two forms, both of which are equally justifiable :

(*a*) The carriage is in motion relative to the embankment.

(*b*) The embankment is in motion relative to the carriage.

In (*a*) the embankment, in (*b*) the carriage, serves as the body of reference in our statement of the motion taking place. If it is simply a question of detecting or of describing the motion involved, it is in principle immaterial to what reference-body we refer the motion. As already mentioned, this is self-evident, but it must not be confused with the much more comprehensive statement called "the principle of relativity," which we have taken as the basis of our investigations.

The principle we have made use of not only maintains that we may equally well choose the carriage or the embankment as our reference-body for the description of any event (for this, too, is self-evident). Our principle rather asserts what follows: If we formulate the general laws of nature as they are obtained from experience, by making use of

(*a*) the embankment as reference-body,

(*b*) the railway carriage as reference-body,

then these general laws of nature (*e.g.* the laws of mechanics or the law of the propagation of light *in vacuo*) have exactly the same form in both cases. This can also be expressed as follows: For the *physical* description of natural processes, neither of the reference-bodies K, K' is unique (lit. "specially marked out") as compared with the other. Unlike the first, this latter statement need not of necessity hold *a priori*; it is not contained in the conceptions of "motion" and "reference-body" and derivable from them; only *experience* can decide as to its correctness or incorrectness.

Up to the present, however, we have by no means maintained the equivalence of *all* bodies of reference K in connection with the formulation of natural laws. Our course was more on the following lines. In the first place, we started out from the assumption that there exists a reference-body K, whose condition of motion is such that the Galileian law holds with respect to it: A particle left to itself and sufficiently far removed from all other particles moves uniformly in a straight line. With reference to K (Galileian reference-body) the laws of nature were to be as simple as possible. But in addition to K, all bodies of reference K' should be given preference in this sense, and they should be exactly equivalent to K for the formulation of natural laws, provided that they are in a state of *uniform rectilinear and non-rotary motion* with respect to K; all these bodies of reference are to be regarded as Galileian reference-bodies. The validity of the principle of relativity was assumed only for these reference-bodies, but not for others (*e.g.* those possessing motion of a different kind). In this sense we speak of the *special* principle of relativity, or special theory of relativity.

In contrast to this we wish to understand by the "general principle of relativity" the following statement: All bodies of reference K, K', etc., are equivalent for the description of natural phenomena (formulation of the general laws of nature), whatever may be their state of motion. But before proceeding farther, it ought to be pointed out that this formulation must be replaced later by a more abstract one, for reasons which will become evident at a later stage.

Since the introduction of the special principle of relativity has been justified, every intellect which strives after generalisation must feel the temptation to venture the step towards the general principle of relativity. But a simple and apparently quite reliable consideration seems to suggest that, for the present at any rate, there is little hope of success in such an attempt. Let us imagine ourselves transferred to our old friend the railway carriage, which is travelling at a uniform rate. As long as it is moving uniformly, the occupant of the carriage is not sensible of its motion, and it is for this reason that he can without reluctance interpret the facts of the case as indicating that the carriage is at rest, but the embankment in motion. Moreover, according to the special principle of relativity, this interpretation is quite justified also from a physical point of view.

If the motion of the carriage is now changed into a non-uniform motion, as for instance by a powerful application of the brakes, then the occupant of the carriage experiences a correspondingly powerful jerk forwards. The retarded motion is manifested in the mechanical behaviour of bodies relative to the person in the railway carriage. The mechanical behaviour is different from that of the case previously considered, and for this reason it would appear to be impossible that the same mechanical laws hold relatively to the non-uniformly moving carriage, as hold with reference to the carriage when at rest or in uniform motion. At all events it is clear that the Galileian law does not hold with respect to the non-uniformly moving carriage. Because of this, we feel compelled at the present juncture to grant a kind of absolute physical reality to non-uniform motion, in opposition to the general principle of relativity. But in what follows we shall soon see that this conclusion cannot be maintained.

XIX

THE GRAVITATIONAL FIELD

"IF we pick up a stone and then let it go, why does it fall to the ground?" The usual answer to this question is: "Because it is attracted by the earth." Modern physics formulates the answer rather differently for the following reason. As a result of the more careful study of electromagnetic phenomena, we have come to regard action at a distance as a process impossible without the intervention of some intermediary medium. If, for instance, a magnet attracts a piece of iron, we cannot be content to regard this as meaning that the magnet acts directly on the iron through the intermediate empty space, but we are constrained to imagine—after the manner of Faraday—that the magnet always calls into being something physically real in the space around it, that something being what we call a "magnetic field." In its turn this magnetic field operates on the piece of iron, so that the latter strives to move towards the magnet. We shall not discuss here the justification for this incidental conception, which is indeed a somewhat arbitrary one. We shall only mention that with its aid electromagnetic phenomena can be theoretically represented much more satisfactorily than without it, and this applies particularly to the transmission of electromagnetic waves.

The effects of gravitation also are regarded in an analogous manner.

The action of the earth on the stone takes place indirectly. The earth produces in its surroundings a gravitational field, which acts on the stone and produces its motion of fall. As we know from experience, the intensity of the action on a body diminishes according to a quite definite law, as we proceed farther and farther away from the earth. From our point of view this means: The law governing the properties of the gravitational field in space must be a perfectly definite one, in order correctly to represent the diminution of gravitational action with the distance from operative bodies. It is something like this: The body (*e.g.* the earth) produces a field in its immediate neighbourhood directly; the intensity and direction of the field at points farther removed from the body are thence determined by the law which governs the properties in space of the gravitational fields themselves.

In contrast to electric and magnetic fields, the gravitational field exhibits a most remarkable property, which is of fundamental importance for what follows. Bodies which are moving under the sole influence of a gravitational field receive an acceleration, *which does not in the least depend either on the material or on the physical state of the body*. For instance, a piece of lead and a piece of wood fall in exactly the same manner in a gravitational field (*in vacuo*), when they start off from rest or with the same initial velocity. This law, which holds most accurately, can be expressed in a different form in the light of the following consideration.

According to Newton's law of motion, we have

(Force) = (inertial mass) × (acceleration),

where the "inertial mass" is a characteristic constant of the accelerated body. If now gravitation is the cause of the acceleration, we then have

$$\text{(Force)} = \text{(gravitational mass)} \times \text{(intensity of the gravitational field)},$$

where the "gravitational mass" is likewise a characteristic constant for the body. From these two relations follows:

$$\text{(acceleration)} = \frac{\text{(gravitational mass)}}{\text{(inertial mass)}} \times \text{(intensity of the gravitational field)}.$$

If now, as we find from experience, the acceleration is to be independent of the nature and the condition of the body and always the same for a given gravitational field, then the ratio of the gravitational to the inertial mass must likewise be the same for all bodies. By a suitable choice of units we can thus make this ratio equal to unity. We then have the following law: The *gravitational* mass of a body is equal to its *inertial* mass.

It is true that this important law had hitherto been recorded in mechanics, but it had not been *interpreted*. A satisfactory interpretation can be obtained only if we recognise the following fact: *The same* quality of a body manifests itself according to circumstances as "inertia" or as "weight" (lit. "heaviness"). In the following section we shall show to what extent this is actually the case, and how this question is connected with the general postulate of relativity.

XX

THE EQUALITY OF INERTIAL AND GRAVITATIONAL MASS AS AN ARGUMENT FOR THE GENERAL POSTULATE OF RELATIVITY

WE imagine a large portion of empty space, so far removed from stars and other appreciable masses, that we have before us approximately the conditions required by the fundamental law of Galilei. It is then possible to choose a Galileian reference-body for this part of space (world), relative to which points at rest remain at rest and points in motion continue permanently in uniform rectilinear motion. As reference-body let us imagine a spacious chest resembling a room with an observer inside who is equipped with apparatus. Gravitation naturally does not exist for this observer. He must fasten himself with strings to the floor, otherwise the slightest impact against the floor will cause him to rise slowly towards the ceiling of the room.

To the middle of the lid of the chest is fixed externally a hook with rope attached, and now a "being" (what kind of a being is immaterial to us) begins pulling at this with a constant force. The chest together with the observer then begin to move "upwards" with a uniformly accelerated motion. In course of time their velocity will reach unheard-of values—provided that we are viewing all this from another reference-body which is not being pulled with a rope.

But how does the man in the chest regard the process ? The acceleration of the chest will be transmitted to him by the reaction of the floor of the chest. He must therefore take up this pressure by means of his legs if he does not wish to be laid out full length on the floor. He is then standing in the chest in exactly the same way as anyone stands in a room of a house on our earth. If he release a body which he previously had in his hand, the acceleration of the chest will no longer be transmitted to this body, and for this reason the body will approach the floor of the chest with an accelerated relative motion. The observer will further convince himself *that the acceleration of the body towards the floor of the chest is always of the same magnitude, whatever kind of body he may happen to use for the experiment.*

Relying on his knowledge of the gravitational field (as it was discussed in the preceding section), the man in the chest will thus come to the conclusion that he and the chest are in a gravitational field which is constant with regard to time. Of course he will be puzzled for a moment as to why the chest does not fall in this gravitational field. Just then, however, he discovers the hook in the middle of the lid of the chest and the rope which is attached to it, and he consequently comes to the conclusion that the chest is suspended at rest in the gravitational field.

Ought we to smile at the man and say that he errs in his conclusion ? I do not believe we ought to if we wish to remain consistent ; we must rather admit that his mode of grasping the situation violates neither reason nor known mechanical laws. Even though it is being accelerated with respect to the "Galileian space" first considered, we can nevertheless regard the chest as being at rest. We have thus good grounds for extending the principle of relativity to include bodies of reference which are accelerated with respect to each other, and as a result we have gained a powerful argument for a generalised postulate of relativity.

We must note carefully that the possibility of this mode of interpretation rests on the fundamental property of the gravitational field of giving all bodies the same acceleration, or, what comes to the same thing, on the law of the equality of inertial and gravitational mass. If this natural law did not exist, the man in the accelerated chest would not be able to interpret the behaviour of the bodies around him on the supposition of a gravitational field, and he would not be justified on the grounds of experience in supposing his reference-body to be "at rest."

Suppose that the man in the chest fixes a rope to the inner side of the lid, and that he attaches a body to the free end of the rope. The result of this will be to stretch the rope so that it will hang "vertically" downwards. If we ask for an opinion of the cause of tension in the rope, the man in the chest will say: "The suspended body experiences a downward force in the gravitational field, and this is neutralised by the tension of the rope; what determines the magnitude of the tension of the rope is the *gravitational mass* of the suspended body." On the other hand, an observer who is poised freely in space will interpret the condition of things thus: "The rope must perforce take part in the accelerated motion of the chest, and it transmits this motion to the body attached to it. The tension of the rope is just large enough to effect the acceleration of the body. That which determines the magnitude of the tension of the rope is the *inertial mass* of the body." Guided by this example, we see that our extension of the principle of relativity implies the *necessity* of the law of the equality of inertial and gravitational mass. Thus we have obtained a physical interpretation of this law.

From our consideration of the accelerated chest we see that a general theory of relativity must yield important results on the laws of gravitation. In point of fact, the systematic pursuit of the general idea of relativity has supplied the laws satisfied by the gravitational field. Before proceeding farther, however, I must warn the reader against a misconception suggested by these considerations. A gravitational field exists for the man in the chest, despite the fact that there was no such field for the co-ordinate system first chosen. Now we might easily suppose that the existence of a gravitational field is always only an *apparent* one. We might also think that, regardless of the kind of gravitational field which may be present, we could always choose another reference-body such that *no* gravitational field exists with reference to it. This is by no means true for all gravitational fields, but only for those of quite special form. It is, for instance, impossible to choose a body of reference such that, as judged from it, the gravitational field of the earth (in its entirety) vanishes.

We can now appreciate why that argument is not convincing, which we brought forward against the general principle of relativity at the end of Section XVIII. It is certainly true that the observer in the railway carriage experiences a jerk forwards as a result of the application of the brake, and that he recognises in this the non-uniformity of motion (retardation) of the carriage. But he is compelled by nobody to refer this jerk to a " real " acceleration (retardation) of the carriage. He might also interpret his experience thus : " My body of reference (the carriage) remains permanently at rest. With reference to it, however, there exists (during the period of application of the brakes) a gravitational field which is directed forwards and which is variable with respect to time. Under the influence of this field, the embankment together with the earth moves non-uniformly in such a manner that their original velocity in the backwards direction is continuously reduced."

XXI

IN WHAT RESPECTS ARE THE FOUNDATIONS OF CLASSICAL MECHANICS AND OF THE SPECIAL THEORY OF RELATIVITY UNSATISFACTORY ?

WE have already stated several times that classical mechanics starts out from the following law: Material particles sufficiently far removed from other material particles continue to move uniformly in a straight line or continue in a state of rest. We have also repeatedly emphasised that this fundamental law can only be valid for bodies of reference *K* which possess certain unique states of motion, and which are in uniform translational motion relative to each other. Relative to other reference-bodies *K* the law is not valid. Both in classical mechanics and in the special theory of relativity we therefore differentiate between reference-bodies *K* relative to which the recognised " laws of nature " can be said to hold, and reference-bodies *K* relative to which these laws do not hold.

But no person whose mode of thought is logical can rest satisfied with this condition of things. He asks: " How does it come that certain reference-bodies (or their states of motion) are given priority over other reference-bodies (or their states of motion) ? *What is the reason for this preference?* In order to show clearly what I mean by this question, I shall make use of a comparison.

I am standing in front of a gas range. Standing alongside of each other on the range are two pans so much alike that one may be mistaken for the other. Both are half full of water. I notice that steam is being emitted continuously from the one pan, but not from the other. I am surprised at this, even if I have never seen either a gas range or a pan before. But if I now notice a luminous something of bluish colour under the first pan but not under the other, I cease to be astonished, even if I have never before seen a gas flame. For I can only say that this bluish something will cause the emission of the steam, or at least *possibly* it may do so. If, however, I notice the bluish something in neither case, and if I observe that the one continuously emits steam whilst the other does not, then I shall remain astonished and dissatisfied until I have discovered some circumstance to which I can attribute the different behaviour of the two pans.

Analogously, I seek in vain for a real something in classical mechanics (or in the special theory of relativity) to which I can attribute the different behaviour of bodies considered with respect to the reference-systems K and K'.[1] Newton saw this objection and attempted to invalidate it, but without success. But E. Mach recognised it most clearly of all, and because of this objection he claimed that mechanics must be

[1] The objection is of importance more especially when the state of motion of the reference-body is of such a nature that it does not require any external agency for its maintenance, *e.g.* in the case when the reference-body is rotating uniformly.

placed on a new basis. It can only be got rid of by means of a physics which is conformable to the general principle of relativity, since the equations of such a theory hold for every body of reference, whatever may be its state of motion.

XXII

A FEW INFERENCES FROM THE GENERAL PRINCIPLE OF RELATIVITY

THE considerations of Section XX show that the general principle of relativity puts us in a position to derive properties of the gravitational field in a purely theoretical manner. Let us suppose, for instance, that we know the space-time " course " for any natural process whatsoever, as regards the manner in which it takes place in the Galileian domain relative to a Galileian body of reference K. By means of purely theoretical operations (*i.e.* simply by calculation) we are then able to find how this known natural process appears, as seen from a reference-body K' which is accelerated relatively to K. But since a gravitational field exists with respect to this new body of reference K', our consideration also teaches us how the gravitational field influences the process studied.

For example, we learn that a body which is in a state of uniform rectilinear motion with respect to K (in accordance with the law of Galilei) is executing an accelerated and in general curvilinear motion with respect to the accelerated reference-body K' (chest). This acceleration or curvature corresponds to the influence on the moving body of the gravitational field prevailing relatively to K'. It is known that a gravitational field influences the movement of bodies in this way, so that our consideration supplies us with nothing essentially new.

However, we obtain a new result of fundamental importance when we carry out the analogous consideration for a ray of light. With respect to the Galileian reference-body K, such a ray of light is transmitted rectilinearly with the velocity c. It can easily be shown that the path of the same ray of light is no longer a straight line when we consider it with reference to the accelerated chest (reference-body K'). From this we conclude, *that, in general, rays of light are propagated curvilinearly in gravitational fields.* In two respects this result is of great importance.

In the first place, it can be compared with the reality. Although a detailed examination of the question shows that the curvature of light rays required by the general theory of relativity is only exceedingly small for the gravitational fields at our disposal in practice, its estimated magnitude for light rays passing the sun at grazing incidence is nevertheless 1·7 seconds of arc. This ought to manifest itself in the following way. As seen from the earth, certain fixed stars appear to be in the neighbourhood of the sun, and are thus capable of observation during a total eclipse of the sun. At such times, these stars ought to appear to be displaced outwards from the sun by an amount indicated above, as compared with their apparent position in the sky when the sun is situated at another part of the heavens. The examination of the correctness or otherwise of this deduction is a problem of the greatest importance, the early solution of which is to be expected of astronomers.[1]

[1] By means of the star photographs of two expeditions equipped by a Joint Committee of the Royal and Royal Astronomical

In the second place our result shows that, according to the general theory of relativity, the law of the constancy of the velocity of light *in vacuo*, which constitutes one of the two fundamental assumptions in the special theory of relativity and to which we have already frequently referred, cannot claim any unlimited validity. A curvature of rays of light can only take place when the velocity of propagation of light varies with position. Now we might think that as a consequence of this, the special theory of relativity and with it the whole theory of relativity would be laid in the dust. But in reality this is not the case. We can only conclude that the special theory of relativity cannot claim an unlimited domain of validity; its results hold only so long as we are able to disregard the influences of gravitational fields on the phenomena (*e.g.* of light).

Since it has often been contended by opponents of the theory of relativity that the special theory of relativity is overthrown by the general theory of relativity, it is perhaps advisable to make the facts of the case clearer by means of an appropriate comparison. Before the development of electrodynamics the laws of electrostatics were looked upon as the laws of electricity. At the present time we know that electric fields can be derived correctly from electrostatic considerations only for the case, which is never strictly realised, in which the electrical masses are quite at rest relatively to each other, and to the co-ordinate system. Should we be justified in saying that for this

Societies, the existence of the deflection of light demanded by theory was confirmed during the solar eclipse of 29th May, 1919. (Cf. Appendix III.)

reason electrostatics is overthrown by the field-equations of Maxwell in electrodynamics? Not in the least. Electrostatics is contained in electrodynamics as a limiting case; the laws of the latter lead directly to those of the former for the case in which the fields are invariable with regard to time. No fairer destiny could be allotted to any physical theory, than that it should of itself point out the way to the introduction of a more comprehensive theory, in which it lives on as a limiting case.

In the example of the transmission of light just dealt with, we have seen that the general theory of relativity enables us to derive theoretically the influence of a gravitational field on the course of natural processes, the laws of which are already known when a gravitational field is absent. But the most attractive problem, to the solution of which the general theory of relativity supplies the key, concerns the investigation of the laws satisfied by the gravitational field itself. Let us consider this for a moment.

We are acquainted with space-time domains which behave (approximately) in a "Galileian" fashion under suitable choice of reference-body, *i.e.* domains in which gravitational fields are absent. If we now refer such a domain to a reference-body K' possessing any kind of motion, then relative to K' there exists a gravitational field which is variable with respect to space and time.[1] The character of this field will of course depend on the motion chosen for K'. According to the general theory of relativity, the general law of the gravitational field must be satisfied for all gravitational fields obtain-

[1] This follows from a generalisation of the discussion in Section XX.

able in this way. Even though by no means all gravitational fields can be produced in this way, yet we may entertain the hope that the general law of gravitation will be derivable from such gravitational fields of a special kind. This hope has been realised in the most beautiful manner. But between the clear vision of this goal and its actual realisation it was necessary to surmount a serious difficulty, and as this lies deep at the root of things, I dare not withhold it from the reader. We require to extend our ideas of the space-time continuum still farther.

XXIII

BEHAVIOUR OF CLOCKS AND MEASURING-RODS ON A ROTATING BODY OF REFERENCE

HITHERTO I have purposely refrained from speaking about the physical interpretation of space- and time-data in the case of the general theory of relativity. As a consequence, I am guilty of a certain slovenliness of treatment, which, as we know from the special theory of relativity, is far from being unimportant and pardonable. It is now high time that we remedy this defect; but I would mention at the outset, that this matter lays no small claims on the patience and on the power of abstraction of the reader.

We start off again from quite special cases, which we have frequently used before. Let us consider a space-time domain in which no gravitational field exists relative to a reference-body K whose state of motion has been suitably chosen. K is then a Galileian reference-body as regards the domain considered, and the results of the special theory of relativity hold relative to K. Let us suppose the same domain referred to a second body of reference K', which is rotating uniformly with respect to K. In order to fix our ideas, we shall imagine K' to be in the form of a plane circular disc, which rotates uniformly in its own plane about its centre. An observer who is sitting eccentrically on the disc K' is sensible of a force which acts outwards in a radial direction, and which would be interpreted as an effect of inertia (centrifugal force) by an observer who was at rest with respect to the original reference-body K. But the observer on the disc may regard his disc as a reference-body which is " at rest " ; on the basis of the general principle of relativity he is justified in doing this. The force acting on himself, and in fact on all other bodies which are at rest relative to the disc, he regards as the effect of a gravitational field. Nevertheless, the space-distribution of this gravitational field is of a kind that would not be possible on Newton's theory of gravitation.[1] But since the observer believes in the general theory of relativity, this does not disturb him; he is quite in the right when he believes that a general law of gravitation can be formulated—a law which not only explains the motion of the stars correctly, but also the field of force experienced by himself.

The observer performs experiments on his circular disc with clocks and measuring-rods. In doing so, it is his intention to arrive at exact definitions for the signification of time- and space-data with reference to the circular disc K', these definitions being based on his observations. What will be his experience in this enterprise ?

To start with, he places one of two identically constructed clocks at the centre of the circular disc, and the other on the edge of the disc, so that they are at rest relative to it. We now ask ourselves whether both clocks go at the same rate from the standpoint of the

[1] The field disappears at the centre of the disc and increases proportionally to the distance from the centre as we proceed outwards.

non-rotating Galileian reference-body K. As judged from this body, the clock at the centre of the disc has no velocity, whereas the clock at the edge of the disc is in motion relative to K in consequence of the rotation. According to a result obtained in Section XII, it follows that the latter clock goes at a rate permanently slower than that of the clock at the centre of the circular disc, *i.e.* as observed from K. It is obvious that the same effect would be noted by an observer whom we will imagine sitting alongside his clock at the centre of the circular disc. Thus on our circular disc, or, to make the case more general, in every gravitational field, a clock will go more quickly or less quickly, according to the position in which the clock is situated (at rest). For this reason it is not possible to obtain a reasonable definition of time with the aid of clocks which are arranged at rest with respect to the body of reference. A similar difficulty presents itself when we attempt to apply our earlier definition of simultaneity in such a case, but I do not wish to go any farther into this question.

Moreover, at this stage the definition of the space co-ordinates also presents insurmountable difficulties. If the observer applies his standard measuring-rod (a rod which is short as compared with the radius of the disc) tangentially to the edge of the disc, then, as judged from the Galileian system, the length of this rod will be less than 1, since, according to Section XII, moving bodies suffer a shortening in the direction of the motion. On the other hand, the measuring-rod will not experience a shortening in length, as judged from K, if it is applied to the disc in the direction of the radius. If, then, the observer first measures the circumference of the disc with his measuring-rod and then the diameter of the disc, on dividing the one by the other, he will not obtain as quotient the familiar number $\pi = 3{\cdot}14 \ldots$, but a larger number,[1] whereas of course, for a disc which is at rest with respect to K, this operation would yield π exactly. This proves that the propositions of Euclidean geometry cannot hold exactly on the rotating disc, nor in general in a gravitational field, at least if we attribute the length 1 to the rod in all positions and in every orientation. Hence the idea of a straight line also loses its meaning. We are therefore not in a position to define exactly the co-ordinates x, y, z relative to the disc by means of the method used in discussing the special theory, and as long as the co-ordinates and times of events have not been defined, we cannot assign an exact meaning to the natural laws in which these occur.

Thus all our previous conclusions based on general relativity would appear to be called in question. In reality we must make a subtle detour in order to be able to apply the postulate of general relativity exactly. I shall prepare the reader for this in the following paragraphs.

[1] Throughout this consideration we have to use the Galileian (non-rotating) system K as reference-body, since we may only assume the validity of the results of the special theory of relativity relative to K (relative to K' a gravitational field prevails).

XXIV

EUCLIDEAN AND NON-EUCLIDEAN CONTINUUM

THE surface of a marble table is spread out in front of me. I can get from any one point on this table to any other point by passing continuously from one point to a " neighbouring " one, and repeating this process a (large) number of times, or, in other words, by going from point to point without executing "jumps." I am sure the reader will appreciate with sufficient clearness what I mean here by " neighbouring " and by " jumps " (if he is not too pedantic). We express this property of the surface by describing the latter as a continuum.

Let us now imagine that a large number of little rods of equal length have been made, their lengths being small compared with the dimensions of the marble slab. When I say they are of equal length, I mean that one can be laid on any other without the ends overlapping. We next lay four of these little rods on the marble slab so that they constitute a quadrilateral figure (a square), the diagonals of which are equally long. To ensure the equality of the diagonals, we make use of a little testing-rod. To this square we add similar ones, each of which has one rod in common with the first. We proceed in like manner with each of these squares until finally the whole marble slab is laid out with squares. The arrangement is such, that each side of a square belongs to two squares and each corner to four squares.

It is a veritable wonder that we can carry out this business without getting into the greatest difficulties. We only need to think of the following. If at any moment three squares meet at a corner, then two sides of the fourth square are already laid, and, as a consequence, the arrangement of the remaining two sides of the square is already completely determined. But I am now no longer able to adjust the quadrilateral so that its diagonals may be equal. If they are equal of their own accord, then this is an especial favour of the marble slab and of the little rods, about which I can only be thankfully surprised. We must needs experience many such surprises if the construction is to be successful.

If everything has really gone smoothly, then I say that the points of the marble slab constitute a Euclidean continuum with respect to the little rod, which has been used as a "distance" (line-interval). By choosing one corner of a square as "origin," I can characterise every other corner of a square with reference to this origin by means of two numbers. I only need state how many rods I must pass over when, starting from the origin, I proceed towards the "right" and then "upwards," in order to arrive at the corner of the square under consideration. These two numbers are then the "Cartesian co-ordinates" of this corner with reference to the "Cartesian co-ordinate system" which is determined by the arrangement of little rods.

By making use of the following modification of this abstract experiment, we recognise that there must also be cases in which the experiment would be unsuccessful. We shall suppose that the rods " expand " by an amount proportional to the increase of temperature. We heat the central part of the marble slab, but not the periphery, in which case two of our little rods can still be brought into coincidence at every position on the table. But our construction of squares must necessarily come into disorder during the heating, because the little rods on the central region of the table expand, whereas those on the outer part do not.

With reference to our little rods—defined as unit lengths—the marble slab is no longer a Euclidean continuum, and we are also no longer in the position of defining Cartesian co-ordinates directly with their aid, since the above construction can no longer be carried out. But since there are other things which are not influenced in a similar manner to the little rods (or perhaps not at all) by the temperature of the table, it is possible quite naturally to maintain the point of view that the marble slab is a "Euclidean continuum." This can be done in a satisfactory manner by making a more subtle stipulation about the measurement or the comparison of lengths.

But if rods of every kind (*i.e.* of every material) were to behave *in the same way* as regards the influence of temperature when they are on the variably heated marble slab, and if we had no other means of detecting the effect of temperature than the geometrical behaviour of our rods in experiments analogous to the one described above, then our best plan would be to assign the distance *one* to two points on the slab, provided that the ends of one of our rods could be made to coincide with these two points; for how else should we define the distance without our proceeding being in the highest measure grossly arbitrary? The method of Cartesian co-ordinates must then be discarded, and replaced by another which does not assume the validity of Euclidean geometry for rigid bodies.[1] The reader will notice that the situation depicted here corresponds to the one brought about by the general postulate of relativity (Section XXIII).

[1] Mathematicians have been confronted with our problem in the following form. If we are given a surface (*e.g.* an ellipsoid) in Euclidean three-dimensional space, then there exists for this surface a two-dimensional geometry, just as much as for a plane surface. Gauss undertook the task of treating this two-dimensional geometry from first principles, without making use of the fact that the surface belongs to a Euclidean continuum of three dimensions. If we imagine constructions to be made with rigid rods *in the surface* (similar to that above with the marble slab), we should find that different laws hold for these from those resulting on the basis of Euclidean plane geometry. The surface is not a Euclidean continuum with respect to the rods, and we cannot define Cartesian co-ordinates *in the surface*. Gauss indicated the principles according to which we can treat the geometrical relationships in the surface, and thus pointed out the way to the method of Riemann of treating multi-dimensional, non-Euclidean *continua*. Thus it is that mathematicians long ago solved the formal problems to which we are led by the general postulate of relativity.

XXV

GAUSSIAN CO-ORDINATES

According to Gauss, this combined analytical and geometrical mode of handling the problem can be arrived at in the following way. We imagine a system of arbitrary curves (see Fig. 4) drawn on the surface of the table. These we designate as u-curves, and we indicate each of them by means of a number. The curves $u=1$, $u=2$ and $u=3$ are drawn in the diagram. Between the curves $u=1$ and $u=2$ we must imagine an infinitely large number to be drawn, all of which correspond to real numbers lying between 1 and 2. We have then a system of u-curves, and this "infinitely dense" system covers the whole surface of the table. These u-curves must not intersect each other, and through each point of the surface one and only one curve must pass. Thus a perfectly definite value of u belongs to every point on the surface of the marble slab. In like manner we imagine a system of v-curves drawn on the surface. These satisfy the same conditions as the u-curves, they are provided with num-

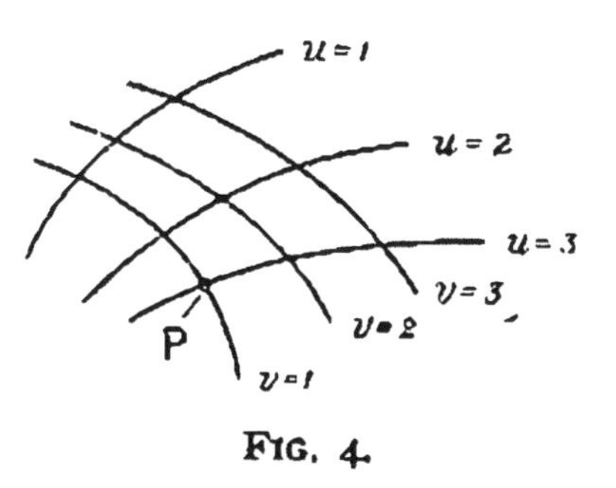

Fig. 4.

bers in a corresponding manner, and they may likewise be of arbitrary shape. It follows that a value of u and a value of v belong to every point on the surface of the table. We call these two numbers the co-ordinates of the surface of the table (Gaussian co-ordinates). For example, the point P in the diagram has the Gaussian co-ordinates $u=3$, $v=1$. Two neighbouring points P and P' on the surface then correspond to the co-ordinates

$$P: \qquad u, v$$
$$P': \qquad u+du, v+dv,$$

where du and dv signify very small numbers. In a similar manner we may indicate the distance (line-interval) between P and P', as measured with a little rod, by means of the very small number ds. Then according to Gauss we have

$$ds^2 = g_{11}du^2 + 2g_{12}dudv + g_{22}\,dv^2,$$

where g_{11}, g_{12}, g_{22}, are magnitudes which depend in a perfectly definite way on u and v. The magnitudes g_{11}, g_{12} and g_{22} determine the behaviour of the rods relative to the u-curves and v-curves, and thus also relative to the surface of the table. For the case in which the points of the surface considered form a Euclidean continuum with reference to the measuring-rods, but only in this case, it is possible to draw the u-curves and v-curves and to attach numbers to them, in such a manner, that we simply have:

$$ds^2 = du^2 + dv^2.$$

Under these conditions, the u-curves and v-curves are straight lines in the sense of Euclidean geometry, and they are perpendicular to each other. Here the Gaussian co-ordinates are simply Cartesian ones. It is clear that Gauss co-ordinates are nothing more than an association of two sets of numbers with the points of the surface considered, of such a nature that numerical values differing very slightly from each other are associated with neighbouring points "in space."

So far, these considerations hold for a continuum of two dimensions. But the Gaussian method can be applied also to a continuum of three, four or more dimensions. If, for instance, a continuum of four dimensions be supposed available, we may represent it in the following way. With every point of the continuum we associate arbitrarily four numbers, x_1, x_2, x_3, x_4, which are known as "co-ordinates." Adjacent points correspond to adjacent values of the co-ordinates. If a distance ds is associated with the adjacent points P and P', this distance being measurable and well-defined from a physical point of view, then the following formula holds:

$$ds^2 = g_{11}dx_1^2 + 2g_{12}dx_1dx_2 \; . \; . \; . \; . \; + g_{44}dx_4^2,$$

where the magnitudes g_{11}, etc., have values which vary with the position in the continuum. Only when the continuum is a Euclidean one is it possible to associate the co-ordinates $x_1 \; . \; . \; x_4$ with the points of the continuum so that we have simply

$$ds^2 = dx_1^2 + dx_2^2 + dx_3^2 + dx_4^2.$$

In this case relations hold in the four-dimensional continuum which are analogous to those holding in our three-dimensional measurements.

However, the Gauss treatment for ds^2 which we have given above is not always possible. It is only possible when sufficiently small regions of the continuum under consideration may be regarded as Euclidean continua.

For example, this obviously holds in the case of the marble slab of the table and local variation of temperature. The temperature is practically constant for a small part of the slab, and thus the geometrical behaviour of the rods is *almost* as it ought to be according to the rules of Euclidean geometry. Hence the imperfections of the construction of squares in the previous section do not show themselves clearly until this construction is extended over a considerable portion of the surface of the table.

We can sum this up as follows: Gauss invented a method for the mathematical treatment of continua in general, in which "size-relations" ("distances" between neighbouring points) are defined. To every point of a continuum are assigned as many numbers (Gaussian co-ordinates) as the continuum has dimensions. This is done in such a way, that only one meaning can be attached to the assignment, and that numbers (Gaussian co-ordinates) which differ by an indefinitely small amount are assigned to adjacent points. The Gaussian co-ordinate system is a logical generalisation of the Cartesian co-ordinate system. It is also applicable to non-Euclidean continua, but only when, with respect to the defined "size" or "distance," small parts of the continuum under consideration behave more nearly like a Euclidean system, the smaller the part of the continuum under our notice.

XXVI

THE SPACE-TIME CONTINUUM OF THE SPECIAL THEORY OF RELATIVITY CONSIDERED AS A EUCLIDEAN CONTINUUM

WE are now in a position to formulate more exactly the idea of Minkowski, which was only vaguely indicated in Section XVII. In accordance with the special theory of relativity, certain co-ordinate systems are given preference for the description of the four-dimensional, space-time continuum. We called these "Galileian co-ordinate systems." For these systems, the four co-ordinates x, y, z, t, which determine an event or—in other words—a point of the four-dimensional continuum, are defined physically in a simple manner, as set forth in detail in the first part of this book. For the transition from one Galileian system to another, which is moving uniformly with reference to the first, the equations of the Lorentz transformation are valid. These last form the basis for the derivation of deductions from the special theory of relativity, and in themselves they are nothing more than the expression of the universal validity of the law of transmission of light for all Galileian systems of reference.

Minkowski found that the Lorentz transformations satisfy the following simple conditions. Let us consider two neighbouring events, the relative position of which in the four-dimensional continuum is given with respect to a Galileian reference-body K by the space co-ordinate differences dx, dy, dz and the time-difference dt. With reference to a second Galileian system we shall suppose that the corresponding differences for these two events are dx', dy', dz', dt'. Then these magnitudes always fulfil the condition [1]

$$dx^2+dy^2+dz^2-c^2dt^2=dx'^2+dy'^2+dz'^2-c^2dt'^2.$$

The validity of the Lorentz transformation follows from this condition. We can express this as follows: The magnitude

$$ds^2=dx^2+dy^2+dz^2-c^2dt^2,$$

which belongs to two adjacent points of the four-dimensional space-time continuum, has the same value for all selected (Galileian) reference-bodies. If we replace x, y, z, $\sqrt{-1}\,ct$, by x_1, x_2, x_3, x_4, we also obtain the result that

$$ds^2=dx_1^2+dx_2^2+dx_3^2+dx_4^2$$

is independent of the choice of the body of reference. We call the magnitude ds the " distance " apart of the two events or four-dimensional points.

Thus, if we choose as time-variable the imaginary variable $\sqrt{-1}\,ct$ instead of the real quantity t, we can regard the space-time continuum—in accordance with the special theory of relativity—as a " Euclidean " four-dimensional continuum, a result which follows from the considerations of the preceding section.

[1] Cf. Appendices I and II. The relations which are derived there for the co-ordinates themselves are valid also for co-ordinate *differences*, and thus also for co-ordinate differentials (indefinitely small differences).

XXVII

THE SPACE-TIME CONTINUUM OF THE GENERAL THEORY OF RELATIVITY IS NOT A EUCLIDEAN CONTINUUM

IN the first part of this book we were able to make use of space-time co-ordinates which allowed of a simple and direct physical interpretation, and which, according to Section XXVI, can be regarded as four-dimensional Cartesian co-ordinates. This was possible on the basis of the law of the constancy of the velocity of light. But according to Section XXI, the general theory of relativity cannot retain this law. On the contrary, we arrived at the result that according to this latter theory the velocity of light must always depend on the co-ordinates when a gravitational field is present. In connection with a specific illustration in Section XXIII, we found that the presence of a gravitational field invalidates the definition of the co-ordinates and the time, which led us to our objective in the special theory of relativity.

In view of the results of these considerations we are led to the conviction that, according to the general principle of relativity, the space-time continuum cannot be regarded as a Euclidean one, but that here we have the general case, corresponding to the marble slab with local variations of temperature, and with which we made acquaintance as an example of a two-dimensional continuum. Just as it was there impossible to construct a Cartesian co-ordinate system from equal rods, so here it is impossible to build up a system (reference-body) from rigid bodies and clocks, which shall be of such a nature that measuring-rods and clocks, arranged rigidly with respect to one another, shall indicate position and time directly. Such was the essence of the difficulty with which we were confronted in Section XXIII.

But the considerations of Sections XXV and XXVI show us the way to surmount this difficulty. We refer the four-dimensional space-time continuum in an arbitrary manner to Gauss co-ordinates. We assign to every point of the continuum (event) four numbers, x_1, x_2, x_3, x_4 (co-ordinates), which have not the least direct physical significance, but only serve the purpose of numbering the points of the continuum in a definite but arbitrary manner. This arrangement does not even need to be of such a kind that we must regard x_1, x_2, x_3 as "space" co-ordinates and x_4 as a "time" co-ordinate.

The reader may think that such a description of the world would be quite inadequate. What does it mean to assign to an event the particular co-ordinates x_1, x_2, x_3, x_4, if in themselves these co-ordinates have no significance? More careful consideration shows, however, that this anxiety is unfounded. Let us consider, for instance, a material point with any kind of motion. If this point had only a momentary existence without duration, then it would be described in space-time by a single system of values x_1, x_2, x_3, x_4. Thus its permanent existence must be characterised by an infinitely large number of such systems of values, the co-ordinate values of which are so close together as to give continuity;

corresponding to the material point, we thus have a (uni-dimensional) line in the four-dimensional continuum. In the same way, any such lines in our continuum correspond to many points in motion. The only statements having regard to these points which can claim a physical existence are in reality the statements about their encounters. In our mathematical treatment, such an encounter is expressed in the fact that the two lines which represent the motions of the points in question have a particular system of co-ordinate values, x_1, x_2, x_3, x_4, in common. After mature consideration the reader will doubtless admit that in reality such encounters constitute the only actual evidence of a time-space nature with which we meet in physical statements.

When we were describing the motion of a material point relative to a body of reference, we stated nothing more than the encounters of this point with particular points of the reference-body. We can also determine the corresponding values of the time by the observation of encounters of the body with clocks, in conjunction with the observation of the encounter of the hands of clocks with particular points on the dials. It is just the same in the case of space-measurements by means of measuring-rods, as a little consideration will show.

The following statements hold generally: Every physical description resolves itself into a number of statements, each of which refers to the space-time coincidence of two events *A* and *B*. In terms of Gaussian co-ordinates, every such statement is expressed by the agreement of their four co-ordinates x_1, x_2, x_3,

Thus in reality, the description of the time-space

continuum by means of Gauss co-ordinates completely replaces the description with the aid of a body of reference, without suffering from the defects of the latter mode of description ; it is not tied down to the Euclidean character of the continuum which has to be represented.

XXVIII

EXACT FORMULATION OF THE GENERAL PRINCIPLE OF RELATIVITY

WE are now in a position to replace the provisional formulation of the general principle of relativity given in Section XVIII by an exact formulation. The form there used, "All bodies of reference K, K', etc., are equivalent for the description of natural phenomena (formulation of the general laws of nature), whatever may be their state of motion," cannot be maintained, because the use of rigid reference-bodies, in the sense of the method followed in the special theory of relativity, is in general not possible in space-time description. The Gauss co-ordinate system has to take the place of the body of reference. The following statement corresponds to the fundamental idea of the general principle of relativity: "*All Gaussian co-ordinate systems are essentially equivalent for the formulation of the general laws of nature.*"

We can state this general principle of relativity in still another form, which renders it yet more clearly intelligible than it is when in the form of the natural extension of the special principle of relativity. According to the special theory of relativity, the equations which express the general laws of nature pass over into equations of the same form when, by making use of the Lorentz transformation, we replace the space-time variables x, y, z, t, of a (Galileian) reference-body K by the space-time variables x', y', z', t', of a new reference-body K'. According to the general theory of relativity, on the other hand, by application of *arbitrary substitutions* of the Gauss variables x_1, x_2, x_3, x_4, the equations must pass over into equations of the same form; for every transformation (not only the Lorentz transformation) corresponds to the transition of one Gauss co-ordinate system into another.

If we desire to adhere to our "old-time" three-dimensional view of things, then we can characterise the development which is being undergone by the fundamental idea of the general theory of relativity as follows: The special theory of relativity has reference to Galileian domains, *i.e.* to those in which no gravitational field exists. In this connection a Galileian reference-body serves as body of reference, *i.e.* a rigid body the state of motion of which is so chosen that the Galileian law of the uniform rectilinear motion of "isolated" material points holds relatively to it.

Certain considerations suggest that we should refer the same Galileian domains to *non-Galileian* reference-bodies also. A gravitational field of a special kind is then present with respect to these bodies (cf. Sections XX and XXIII).

In gravitational fields there are no such things as rigid bodies with Euclidean properties; thus the fictitious rigid body of reference is of no avail in the general theory of relativity. The motion of clocks is also influenced by gravitational fields, and in such a way that a physical definition of time which is made directly with the aid of clocks has by no means the same degree of plausibility as in the special theory of relativity.

For this reason non-rigid reference-bodies are used, which are as a whole not only moving in any way whatsoever, but which also suffer alterations in form *ad lib.* during their motion. Clocks, for which the law of motion is of any kind, however irregular, serve for the definition of time. We have to imagine each of these clocks fixed at a point on the non-rigid reference-body. These clocks satisfy only the one condition, that the "readings" which are observed simultaneously on adjacent clocks (in space) differ from each other by an indefinitely small amount. This non-rigid reference-body, which might appropriately be termed a "reference-mollusk," is in the main equivalent to a Gaussian four-dimensional co-ordinate system chosen arbitrarily. That which gives the "mollusk" a certain comprehensibleness as compared with the Gauss co-ordinate system is the (really unjustified) formal retention of the separate existence of the space co-ordinates as opposed to the time co-ordinate. Every point on the mollusk is treated as a space-point, and every material point which is at rest relatively to it as at rest, so long as the mollusk is considered as reference-body. The general principle of relativity requires that all these mollusks can be used as reference-bodies with equal right and equal success in the formulation of the general laws of nature; the laws themselves must be quite independent of the choice of mollusk.

The great power possessed by the general principle of relativity lies in the comprehensive limitation which is imposed on the laws of nature in consequence of what we have seen above.

XXIX

THE SOLUTION OF THE PROBLEM OF GRAVITATION ON THE BASIS OF THE GENERAL PRINCIPLE OF RELATIVITY

IF the reader has followed all our previous considerations, he will have no further difficulty in understanding the methods leading to the solution of the problem of gravitation.

We start off from a consideration of a Galileian domain, *i.e.* a domain in which there is no gravitational field relative to the Galileian reference-body K. The behaviour of measuring-rods and clocks with reference to K is known from the special theory of relativity, likewise the behaviour of "isolated" material points; the latter move uniformly and in straight lines.

Now let us refer this domain to a random Gauss coordinate system or to a "mollusk" as reference-body K'. Then with respect to K' there is a gravitational field G (of a particular kind). We learn the behaviour of measuring-rods and clocks and also of freely-moving material points with reference to K' simply by mathematical transformation. We interpret this behaviour as the behaviour of measuring-rods, clocks and material points under the influence of the gravitational field G. Hereupon we introduce a hypothesis: that the influence of the gravitational field on measuring-rods, clocks and freely-moving material points continues to take place according to the same laws, even in the case when the prevailing gravitational field is *not* derivable from the Galileian special case, simply by means of a transformation of co-ordinates.

The next step is to investigate the space-time behaviour of the gravitational field G, which was derived from the Galileian special case simply by transformation of the co-ordinates. This behaviour is formulated in a law, which is always valid, no matter how the reference-body (mollusk) used in the description may be chosen.

This law is not yet the *general* law of the gravitational field, since the gravitational field under consideration is of a special kind. In order to find out the general law-of-field of gravitation we still require to obtain a generalisation of the law as found above. This can be obtained without caprice, however, by taking into consideration the following demands:

(*a*) The required generalisation must likewise satisfy the general postulate of relativity.

(*b*) If there is any matter in the domain under consideration, only its inertial mass, and thus according to Section XV only its energy is of importance for its effect in exciting a field.

(*c*) Gravitational field and matter together must satisfy the law of the conservation of energy (and of impulse).

Finally, the general principle of relativity permits us to determine the influence of the gravitational field on the course of all those processes which take place according to known laws when a gravitational field is absent, *i.e.* which have already been fitted into the frame of the special theory of relativity. In this connection we proceed in principle according to the method which has already been explained for measuring-rods, clocks and freely-moving material points.

The theory of gravitation derived in this way from the general postulate of relativity excels not only in its beauty; nor in removing the defect attaching to classical mechanics which was brought to light in Section XXI; nor in interpreting the empirical law of the equality of inertial and gravitational mass; but it has also already explained a result of observation in astronomy, against which classical mechanics is powerless.

If we confine the application of the theory to the case where the gravitational fields can be regarded as being weak, and in which all masses move with respect to the co-ordinate system with velocities which are small compared with the velocity of light, we then obtain as a first approximation the Newtonian theory. Thus the latter theory is obtained here without any particular assumption, whereas Newton had to introduce the hypothesis that the force of attraction between mutually attracting material points is inversely proportional to the square of the distance between them. If we increase the accuracy of the calculation, deviations from the theory of Newton make their appearance, practically all of which must nevertheless escape the test of observation owing to their smallness.

We must draw attention here to one of these deviations. According to Newton's theory, a planet moves round the sun in an ellipse, which would permanently maintain its position with respect to the fixed stars, if we could disregard the motion of the fixed stars themselves and the action of the other planets under consideration. Thus, if we correct the observed motion of the planets for these two influences, and if Newton's theory be strictly correct, we ought to obtain for the orbit of the planet an ellipse, which is fixed with reference to the fixed stars. This deduction, which can be tested with great accuracy, has been confirmed for all the planets save one, with the precision that is capable of being obtained by the delicacy of observation attainable at the present time. The sole exception is Mercury, the planet which lies nearest the sun. Since the time of Leverrier, it has been known that the ellipse corresponding to the orbit of Mercury, after it has been corrected for the influences mentioned above, is not stationary with respect to the fixed stars, but that it rotates exceedingly slowly in the plane of the orbit and in the sense of the orbital motion. The value obtained for this rotary movement of the orbital ellipse was 43 seconds of arc per century, an amount ensured to be correct to within a few seconds of arc. This effect can be explained by means of classical mechanics only on the assumption of hypotheses which have little probability, and which were devised solely for this purpose.

On the basis of the general theory of relativity, it is found that the ellipse of every planet round the sun must necessarily rotate in the manner indicated above; that for all the planets, with the exception of Mercury, this rotation is too small to be detected with the delicacy of observation possible at the present time; but that in the case of Mercury it must amount to 43 seconds of arc per century, a result which is strictly in agreement with observation.

Apart from this one, it has hitherto been possible to make only two deductions from the theory which admit of being tested by observation, to wit, the curvature of light rays by the gravitational field of the sun,[1] and a displacement of the spectral lines of light reaching us from large stars, as compared with the corresponding lines for light produced in an analogous manner terrestrially (*i.e.* by the same kind of molecule). I do not doubt that these deductions from the theory will be confirmed also.

[1] Observed by Eddington and others in 1919. (Cf. Appendix III.)

PART III

CONSIDERATIONS ON THE UNIVERSE AS A WHOLE

XXX

COSMOLOGICAL DIFFICULTIES OF NEWTON'S THEORY

APART from the difficulty discussed in Section XXI, there is a second fundamental difficulty attending classical celestial mechanics, which, to the best of my knowledge, was first discussed in detail by the astronomer Seeliger. If we ponder over the question as to how the universe, considered as a whole, is to be regarded, the first answer that suggests itself to us is surely this: As regards space (and time) the universe is infinite. There are stars everywhere, so that the density of matter, although very variable in detail, is nevertheless on the average everywhere the same. In other words: However far we might travel through space, we should find everywhere an attenuated swarm of fixed stars of approximately the same kind and density.

This view is not in harmony with the theory of Newton. The latter theory rather requires that the universe should have a kind of centre in which the density of the stars is a maximum, and that as we proceed outwards from this centre the group-density of the stars should diminish, until finally, at great distances, it is succeeded by an infinite region of emptiness. The stellar universe ought to be a finite island in the infinite ocean of space.[1]

This conception is in itself not very satisfactory. It is still less satisfactory because it leads to the result that the light emitted by the stars and also individual stars of the stellar system are perpetually passing out into infinite space, never to return, and without ever again coming into interaction with other objects of nature. Such a finite material universe would be destined to become gradually but systematically impoverished.

In order to escape this dilemma, Seeliger suggested a modification of Newton's law, in which he assumes that for great distances the force of attraction between two masses diminishes more rapidly than would result from the inverse square law. In this way it is possible for the mean density of matter to be constant everywhere, even to infinity, without infinitely large gravitational fields being produced. We thus free ourselves from the

[1] *Proof*—According to the theory of Newton, the number of "lines of force" which come from infinity and terminate in a mass m is proportional to the mass m. If, on the average, the mass-density ρ_0 is constant throughout the universe, then a sphere of volume V will enclose the average mass $\rho_0 V$. Thus the number of lines of force passing through the surface F of the sphere into its interior is proportional to $\rho_0 V$. For unit area of the surface of the sphere the number of lines of force which enters the sphere is thus proportional to $\rho_0 \frac{V}{F}$ or to $\rho_0 R$. Hence the intensity of the field at the surface would ultimately become infinite with increasing radius R of the sphere, which is impossible.

distasteful conception that the material universe ought to possess something of the nature of a centre. Of course we purchase our emancipation from the fundamental difficulties mentioned, at the cost of a modification and complication of Newton's law which has neither empirical nor theoretical foundation. We can imagine innumerable laws which would serve the same purpose, without our being able to state a reason why one of them is to be preferred to the others; for any one of these laws would be founded just as little on more general theoretical principles as is the law of Newton.

XXXI

THE POSSIBILITY OF A "FINITE" AND YET "UNBOUNDED" UNIVERSE

BUT speculations on the structure of the universe also move in quite another direction. The development of non-Euclidean geometry led to the recognition of the fact, that we can cast doubt on the *infiniteness* of our space without coming into conflict with the laws of thought or with experience (Riemann, Helmholtz). These questions have already been treated in detail and with unsurpassable lucidity by Helmholtz and Poincaré, whereas I can only touch on them briefly here.

In the first place, we imagine an existence in two-dimensional space. Flat beings with flat implements, and in particular flat rigid measuring-rods, are free to move in a *plane*. For them nothing exists outside of this plane: that which they observe to happen to themselves and to their flat "things" is the all-inclusive reality of their plane. In particular, the constructions of plane Euclidean geometry can be carried out by means of the rods, *e.g.* the lattice construction, considered in Section XXIV. In contrast to ours, the universe of these beings is two-dimensional; but, like ours, it extends to infinity. In their universe there is room for an infinite number of identical squares made up of rods, *i.e.* its volume (surface) is infinite. If these beings say their universe is "plane," there is sense in the statement, because they mean that they can perform the constructions of plane Euclidean geometry with their rods. In this connection the individual rods always represent the same distance, independently of their position.

Let us consider now a second two-dimensional existence, but this time on a spherical surface instead of on a plane. The flat beings with their measuring-rods and other objects fit exactly on this surface and they are unable to leave it. Their whole universe of observation extends exclusively over the surface of the sphere. Are these beings able to regard the geometry of their universe as being plane geometry and their rods withal as the realisation of "distance"? They cannot do this. For if they attempt to realise a straight line, they will obtain a curve, which we "three-dimensional beings" designate as a great circle, *i.e.* a self-contained line of definite finite length, which can be measured up by means of a measuring-rod. Similarly, this universe has a finite area that can be compared with the area of a square constructed with rods. The great charm resulting from this consideration lies in the recognition of the fact that *the universe of these beings is finite and yet has no limits.*

But the spherical-surface beings do not need to go on a world-tour in order to perceive that they are not living in a Euclidean universe. They can convince themselves of this on every part of their "world," provided they do not use too small a piece of it. Starting from a point, they draw "straight lines" (arcs of circles as judged in three-dimensional space) of equal length in all directions. They will call the line joining the free ends of these lines a "circle." For a plane surface, the ratio of the circumference of a circle to its diameter, both lengths being measured with the same rod, is, according to Euclidean geometry of the plane, equal to a constant value π, which is independent of the diameter of the circle. On their spherical surface our flat beings would find for this ratio the value

$$\pi \frac{\sin\left(\frac{r}{R}\right)}{\left(\frac{r}{R}\right)},$$

i.e. a smaller value than π, the difference being the more considerable, the greater is the radius of the circle in comparison with the radius R of the "world-sphere." By means of this relation the spherical beings can determine the radius of their universe ("world"), even when only a relatively small part of their world-sphere is available for their measurements. But if this part is very small indeed, they will no longer be able to demonstrate that they are on a spherical "world" and not on a Euclidean plane, for a small part of a spherical surface differs only slightly from a piece of a plane of the same size.

Thus if the spherical-surface beings are living on a planet of which the solar system occupies only a negligibly small part of the spherical universe, they have no means of determining whether they are living in a finite or in an infinite universe, because the "piece of universe" to which they have access is in both cases practically plane, or Euclidean. It follows directly from this discussion, that for our sphere-beings the circumference of a circle first increases with the radius until the "circumference of the universe" is reached, and that it thenceforward gradually decreases to zero for still further increasing values of the radius. During this process the area of the circle continues to increase more and more, until finally it becomes equal to the total area of the whole "world-sphere."

Perhaps the reader will wonder why we have placed our "beings" on a sphere rather than on another closed surface. But this choice has its justification in the fact that, of all closed surfaces, the sphere is unique in possessing the property that all points on it are equivalent. I admit that the ratio of the circumference c of a circle to its radius r depends on r, but for a given value of r it is the same for all points of the "world-sphere"; in other words, the "world-sphere" is a "surface of constant curvature."

To this two-dimensional sphere-universe there is a three-dimensional analogy, namely, the three-dimensional spherical space which was discovered by Riemann. Its points are likewise all equivalent. It possesses a finite volume, which is determined by its "radius" ($2\pi^2R^3$). Is it possible to imagine a spherical space? To imagine a space means nothing else than that we imagine an epitome of our "space" experience, *i.e.* of experience that we can have in the movement of "rigid" bodies. In this sense we *can* imagine a spherical space.

Suppose we draw lines or stretch strings in all directions from a point, and mark off from each of these the distance r with a measuring-rod. All the free end-points of these lengths lie on a spherical surface. We can specially measure up the area (F) of this surface by means of a square made up of measuring-rods. If the universe is Euclidean, then $F=4\pi r^2$; if it is spherical, then F is always less than $4\pi r^2$. With increasing values of r, F increases from zero up to a maximum value which is determined by the "world-radius," but for still further increasing values of r, the area gradually diminishes to zero. At first, the straight lines which radiate from the starting point diverge farther and farther from one another, but later they approach each other, and finally they run together again at a "counter-point" to the starting point. Under such conditions they have traversed the whole spherical space. It is easily seen that the three-dimensional spherical space is quite analogous to the two-dimensional spherical surface. It is finite (*i.e.* of finite volume), and has no bounds.

It may be mentioned that there is yet another kind of curved space: "elliptical space." It can be regarded as a curved space in which the two "counter-points" are identical (indistinguishable from each other). An elliptical universe can thus be considered to some extent as a curved universe possessing central symmetry.

It follows from what has been said, that closed spaces without limits are conceivable. From amongst these, the spherical space (and the elliptical) excels in its simplicity, since all points on it are equivalent. As a result of this discussion, a most interesting question arises for astronomers and physicists, and that is whether the universe in which we live is infinite, or whether it is finite in the manner of the spherical universe. Our experience is far from being sufficient to enable us to answer this question. But the general theory of relativity permits of our answering it with a moderate degree of certainty, and in this connection the difficulty mentioned in Section XXX finds its solution.

XXXII

THE STRUCTURE OF SPACE ACCORDING TO THE GENERAL THEORY OF RELATIVITY

ACCORDING to the general theory of relativity, the geometrical properties of space are not independent, but they are determined by matter. Thus we can draw conclusions about the geometrical structure of the universe only if we base our considerations on the state of the matter as being something that is known. We know from experience that, for a suitably chosen co-ordinate system, the velocities of the stars are small as compared with the velocity of transmission of light. We can thus as a rough approximation arrive at a conclusion as to the nature of the universe as a whole, if we treat the matter as being at rest.

We already know from our previous discussion that the behaviour of measuring-rods and clocks is influenced by gravitational fields, *i.e.* by the distribution of matter. This in itself is sufficient to exclude the possibility of the exact validity of Euclidean geometry in our universe. But it is conceivable that our universe differs only slightly from a Euclidean one, and this notion seems all the more probable, since calculations show that the metrics of surrounding space is influenced only to an exceedingly small extent by masses even of the magnitude of our sun. We might imagine that, as regards geometry, our universe behaves analogously to a surface which is irregularly curved in its individual parts, but which nowhere departs appreciably from a plane: something like the rippled surface of a lake. Such a universe might fittingly be called a quasi-Euclidean universe. As regards its space it would be infinite. But calculation shows that in a quasi-Euclidean universe the average density of matter would necessarily be *nil*. Thus such a universe could not be inhabited by matter everywhere; it would present to us that unsatisfactory picture which we portrayed in Section XXX.

If we are to have in the universe an average density of matter which differs from zero, however small may be that difference, then the universe cannot be quasi-Euclidean. On the contrary, the results of calculation indicate that if matter be distributed uniformly, the universe would necessarily be spherical (or elliptical). Since in reality the detailed distribution of matter is not uniform, the real universe will deviate in individual parts from the spherical, *i.e.* the universe will be quasi-spherical. But it will be necessarily finite. In fact, the theory supplies us with a simple connection[1] between the space-expanse of the universe and the average density of matter in it.

[1] For the "radius" R of the universe we obtain the equation

$$R^2 = \frac{2}{\kappa\rho}.$$

The use of the C.G.S. system in this equation gives $\frac{2}{\kappa} = 1{\cdot}08 \,.\, 10^{27}$; is the average density of the matter.

APPENDIX I

SIMPLE DERIVATION OF THE LORENTZ TRANSFORMATION [SUPPLEMENTARY TO SECTION XI]

FOR the relative orientation of the co-ordinate systems indicated in Fig. 2, the x-axes of both systems permanently coincide. In the present case we can divide the problem into parts by considering first only events which are localised on the x-axis. Any such event is represented with respect to the co-ordinate system K by the abscissa x and the time t, and with respect to the system K' by the abscissa x' and the time t'. We require to find x' and t' when x and t are given.

A light-signal, which is proceeding along the positive axis of x, is transmitted according to the equation

$$x = ct$$

or

$$x - ct = 0 \quad . \quad . \quad . \quad (1).$$

Since the same light-signal has to be transmitted relative to K' with the velocity c, the propagation relative to the system K' will be represented by the analogous formula

$$x' - ct' = 0 \quad . \quad . \quad . \quad (2)$$

Those space-time points (events) which satisfy (1) must

also satisfy (2). Obviously this will be the case when the relation

$$(x'-ct')=\lambda(x-ct) \quad . \quad . \quad . \quad (3).$$

is fulfilled in general, where λ indicates a constant; for, according to (3), the disappearance of $(x-ct)$ involves the disappearance of $(x'-ct')$.

If we apply quite similar considerations to light rays which are being transmitted along the negative x-axis, we obtain the condition

$$(x'+ct')=\mu(x+ct) \quad . \quad . \quad . \quad (4).$$

By adding (or subtracting) equations (3) and (4), and introducing for convenience the constants a and b in place of the constants λ and μ, where

$$a=\frac{\lambda+\mu}{2}$$

and

$$b=\frac{\lambda-\mu}{2},$$

we obtain the equations

$$\left.\begin{aligned} x' &= ax-bct \\ ct' &= act-bx \end{aligned}\right\} \quad . \quad . \quad . \quad (5).$$

We should thus have the solution of our problem, if the constants a and b were known. These result from the following discussion.

For the origin of K' we have permanently $x'=0$, and hence according to the first of the equations (5)

$$x=\frac{bc}{a}t.$$

If we call v the velocity with which the origin of K' is moving relative to K, we then have

$$v=\frac{bc}{a} \quad . \quad . \quad . \quad . \quad (6).$$

The same value v can be obtained from equation (5), if we calculate the velocity of another point of K' relative to K, or the velocity (directed towards the negative x-axis) of a point of K with respect to K'. In short, we can designate v as the relative velocity of the two systems.

Furthermore, the principle of relativity teaches us that, as judged from K, the length of a unit measuring-rod which is at rest with reference to K' must be exactly the same as the length, as judged from K', of a unit measuring-rod which is at rest relative to K. In order to see how the points of the x'-axis appear as viewed from K, we only require to take a "snapshot" of K' from K; this means that we have to insert a particular value of t (time of K), e.g. $t=0$. For this value of t we then obtain from the first of the equations (5)

$$x'=ax.$$

Two points of the x'-axis which are separated by the distance $\Delta x'=1$ when measured in the K' system are thus separated in our instantaneous photograph by the distance

$$\Delta x=\frac{1}{a} \quad . \quad . \quad . \quad (7).$$

But if the snapshot be taken from $K'(t'=0)$, and if we eliminate t from the equations (5), taking into account the expression (6), we obtain

$$x'=a\left(1-\frac{v^2}{c^2}\right)x.$$

From this we conclude that two points on the x-axis and separated by the distance 1 (relative to K) will be represented on our snapshot by the distance

$$\Delta x'=a\left(1-\frac{v^2}{c^2}\right) \quad (7a).$$

But from what has been said, the two snapshots must be identical; hence Δx in (7) must be equal to $\Delta x'$ in (7*a*), so that we obtain

$$a^2 = \frac{1}{1 - \frac{v^2}{c^2}} \quad . \quad . \quad . \quad (7b).$$

The equations (6) and (7*b*) determine the constants *a* and *b*. By inserting the values of these constants in (5), we obtain the first and the fourth of the equations given in Section XI.

$$\left.\begin{aligned} x' &= \frac{x - vt}{\sqrt{1 - \frac{v^2}{c^2}}} \\ t' &= \frac{t - \frac{v}{c^2}x}{\sqrt{1 - \frac{v^2}{c^2}}} \end{aligned}\right\} \quad . \quad . \quad . \quad (8).$$

Thus we have obtained the Lorentz transformation for events on the *x*-axis. It satisfies the condition

$$x'^2 - c^2t'^2 = x^2 - c^2t^2 \quad . \quad . \quad (8a).$$

The extension of this result, to include events which take place outside the *x*-axis, is obtained by retaining equations (8) and supplementing them by the relations

$$\left.\begin{aligned} y' &= y \\ z' &= z \end{aligned}\right\} \quad . \quad . \quad . \quad . \quad (9).$$

In this way we satisfy the postulate of the constancy of the velocity of light *in vacuo* for rays of light of arbitrary direction, both for the system *K* and for the system *K'*. This may be shown in the following manner.

We suppose a light-signal sent out from the origin of *K* at the time $t=0$. It will be propagated according to the equation

$$r = \sqrt{x^2 + y^2 + z^2} = ct,$$

or, if we square this equation, according to the equation

$$x^2+y^2+z^2-c^2t^2=0 \quad . \quad . \quad (10).$$

It is required by the law of propagation of light, in conjunction with the postulate of relativity, that the transmission of the signal in question should take place —as judged from K'—in accordance with the corresponding formula

$$r'=ct',$$

or,

$$x'^2+y'^2+z'^2-c^2t'^2=0 \quad . \quad . \quad (10a).$$

In order that equation (10*a*) may be a consequence of equation (10), we must have

$$x'^2+y'^2+z'^2-c^2t'^2=\sigma(x^2+y^2+z^2-c^2t^2) \quad (11).$$

Since equation (8*a*) must hold for points on the x-axis, we thus have $\sigma=1$. It is easily seen that the Lorentz transformation really satisfies equation (11) for $\sigma=1$; for (11) is a consequence of (8*a*) and (9), and hence also of (8) and (9). We have thus derived the Lorentz transformation.

The Lorentz transformation represented by (8) and (9) still requires to be generalised. Obviously it is immaterial whether the axes of K' be chosen so that they are spatially parallel to those of K. It is also not essential that the velocity of translation of K' with respect to K should be in the direction of the x-axis. A simple consideration shows that we are able to construct the Lorentz transformation in this general sense from two kinds of transformations, viz. from Lorentz transformations in the special sense and from purely spatial transformations, which corresponds to the replacement of the rectangular co-ordinate system by a new system with its axes pointing in other directions.

Mathematically, we can characterise the generalised Lorentz transformation thus:

It expresses x', y', z', t', in terms of linear homogeneous functions of x, y, z, t, of such a kind that the relation

$$x'^2+y'^2+z'^2-c^2t'^2=x^2+y^2+z^2-c^2t^2 \quad . \quad . \quad (11a).$$

is satisfied identically. That is to say: If we substitute their expressions in x, y, z, t, in place of x', y', z', t', on the left-hand side, then the left-hand side of (11a) agrees with the right-hand side.

APPENDIX II

MINKOWSKI'S FOUR-DIMENSIONAL SPACE ("WORLD") [SUPPLEMENTARY TO SECTION XVII]

WE can characterise the Lorentz transformation still more simply if we introduce the imaginary $\sqrt{-1}\,.\,ct$ in place of t, as time-variable. If, in accordance with this, we insert

$$\begin{aligned} x_1 &= x \\ x_2 &= y \\ x_3 &= z \\ x_4 &= \sqrt{-1}\,.\,ct, \end{aligned}$$

and similarly for the accented system K', then the condition which is identically satisfied by the transformation can be expressed thus :

$$x_1'^2 + x_2'^2 + x_3'^2 + x_4'^2 = x_1^2 + x_2^2 + x_3^2 + x_4^2 \quad . \qquad (12).$$

That is, by the afore-mentioned choice of "coordinates," (11*a*) is transformed into this equation.

We see from (12) that the imaginary time co-ordinate x_4 enters into the condition of transformation in exactly the same way as the space co-ordinates x_1, x_2, x_3. It is due to this fact that, according to the theory of relativity, the " time " x_4 enters into natural laws in the same form as the space co-ordinates x_1, x_2, x_3.

A four-dimensional continuum described by the " co-ordinates " x_1, x_2, x_3, x_4, was called " world " by Minkowski, who also termed a point-event a " world-point." From a " happening " in three-dimensional space, physics becomes, as it were, an " existence " in the four-dimensional " world."

This four-dimensional " world " bears a close similarity to the three-dimensional " space " of (Euclidean) analytical geometry. If we introduce into the latter a new Cartesian co-ordinate system (x'_1, x'_2, x'_3) with the same origin, then x'_1, x'_2, x'_3, are linear homogeneous functions of x_1, x_2, x_3, which identically satisfy the equation

$$x_1'^2+x_2'^2+x_3'^2=x_1^2+x_2^2+x_3^2.$$

The analogy with (12) is a complete one. We can regard Minkowski's " world " in a formal manner as a four-dimensional Euclidean space (with imaginary time co-ordinate) ; the Lorentz transformation corresponds to a " rotation " of the co-ordinate system in the four-dimensional " world."

APPENDIX III

THE EXPERIMENTAL CONFIRMATION OF THE GENERAL THEORY OF RELATIVITY

FROM a systematic theoretical point of view, we may imagine the process of evolution of an empirical science to be a continuous process of induction. Theories are evolved and are expressed in short compass as statements of a large number of individual observations in the form of empirical laws, from which the general laws can be ascertained by comparison. Regarded in this way, the development of a science bears some resemblance to the compilation of a classified catalogue. It is, as it were, a purely empirical enterprise.

But this point of view by no means embraces the whole of the actual process; for it slurs over the important part played by intuition and deductive thought in the development of an exact science. As soon as a science has emerged from its initial stages, theoretical advances are no longer achieved merely by a process of arrangement. Guided by empirical data, the investigator rather develops a system of thought which, in general, is built up logically from a small number of fundamental assumptions, the so-called axioms. We call such a system of thought a *theory*. The theory finds the justification for its existence in the fact that it correlates a large number of single observations, and it is just here that the " truth " of the theory lies.

Corresponding to the same complex of empirical data, there may be several theories, which differ from one another to a considerable extent. But as regards the deductions from the theories which are capable of being tested, the agreement between the theories may be so complete, that it becomes difficult to find such deductions in which the two theories differ from each other. As an example, a case of general interest is available in the province of biology, in the Darwinian theory of the development of species by selection in the struggle for existence, and in the theory of development which is based on the hypothesis of the hereditary transmission of acquired characters.

We have another instance of far-reaching agreement between the deductions from two theories in Newtonian mechanics on the one hand, and the general theory of relativity on the other. This agreement goes so far, that up to the present we have been able to find only a few deductions from the general theory of relativity which are capable of investigation, and to which the physics of pre-relativity days does not also lead, and this despite the profound difference in the fundamental assumptions of the two theories. In what follows, we shall again consider these important deductions, and we shall also discuss the empirical evidence appertaining to them which has hitherto been obtained.

(*a*) Motion of the Perihelion of Mercury

According to Newtonian mechanics and Newton's law of gravitation, a planet which is revolving round the sun would describe an ellipse round the latter, or, more correctly, round the common centre of gravity of the sun and the planet. In such a system, the sun, or the common centre of gravity, lies in one of the foci of the orbital ellipse in such a manner that, in the course of a planet-year, the distance sun–planet grows from a minimum to a maximum, and then decreases again to a minimum. If instead of Newton's law we insert a somewhat different law of attraction into the calculation, we find that, according to this new law, the motion would still take place in such a manner that the distance sun–planet exhibits periodic variations; but in this case the angle described by the line joining sun and planet during such a period (from perihelion—closest proximity to the sun—to perihelion) would differ from 360°. The line of the orbit would not then be a closed one, but in the course of time it would fill up an annular part of the orbital plane, viz. between the circle of least and the circle of greatest distance of the planet from the sun.

According also to the general theory of relativity, which differs of course from the theory of Newton, a small variation from the Newton-Kepler motion of a planet in its orbit should take place, and in such a way, that the angle described by the radius sun–planet between one perihelion and the next should exceed that corresponding to one complete revolution by an amount given by

$$+\frac{24\pi^3 a^2}{T^2 c^2 (1-e^2)}.$$

(*N.B.*—One complete revolution corresponds to the angle 2π in the absolute angular measure customary in physics, and the above expression gives the amount by which the radius sun–planet exceeds this angle during the interval between one perihelion and the next.) In this expression a represents the major semi-axis of the ellipse, e its eccentricity, c the velocity of light, and T the period of revolution of the planet. Our result may also be stated as follows: According to the general theory of relativity, the major axis of the ellipse rotates round the sun in the same sense as the orbital motion of the planet. Theory requires that this rotation should amount to 43 seconds of arc per century for the planet Mercury, but for the other planets of our solar system its magnitude should be so small that it would necessarily escape detection.[1]

In point of fact, astronomers have found that the theory of Newton does not suffice to calculate the observed motion of Mercury with an exactness corresponding to that of the delicacy of observation attainable at the present time. After taking account of all the disturbing influences exerted on Mercury by the remaining planets, it was found (Leverrier—1859—and Newcomb—1895) that an unexplained perihelial movement of the orbit of Mercury remained over, the amount of which does not differ sensibly from the abovementioned +43 seconds of arc per century. The uncertainty of the empirical result amounts to a few seconds only.

(b) Deflection of Light by a Gravitational Field

In Section XXII it has been already mentioned that,

[1] Especially since the next planet Venus has an orbit that is almost an exact circle, which makes it more difficult to locate the perihelion with precision.

according to the general theory of relativity, a ray of light will experience a curvature of its path when passing through a gravitational field, this curvature being similar to that experienced by the path of a body which is projected through a gravitational field. As a result of this theory, we should expect that a ray of light which is passing close to a heavenly body would be deviated towards the latter. For a ray of light which passes the sun at a distance of Δ sun-radii from its centre, the angle of deflection (α) should amount to

$$\alpha = \frac{1 \cdot 7 \text{ seconds of arc}}{\Delta}$$

It may be added that, according to the theory, half of this deflection is produced by the Newtonian field of attraction of the sun, and the other half by the geometrical modification ("curvature") of space caused by the sun.

This result admits of an experimental test by means of the photographic registration of stars during a total eclipse of the sun. The only reason why we must wait for a total eclipse is because at every other time the atmosphere is so strongly illuminated by the light from the sun that the stars situated near the sun's disc are invisible. The predicted effect can be seen clearly from the accompanying diagram. If the sun (S) were not present, a star which is practically infinitely distant would be seen in the direction D_1, as observed from the earth. But as a consequence of the

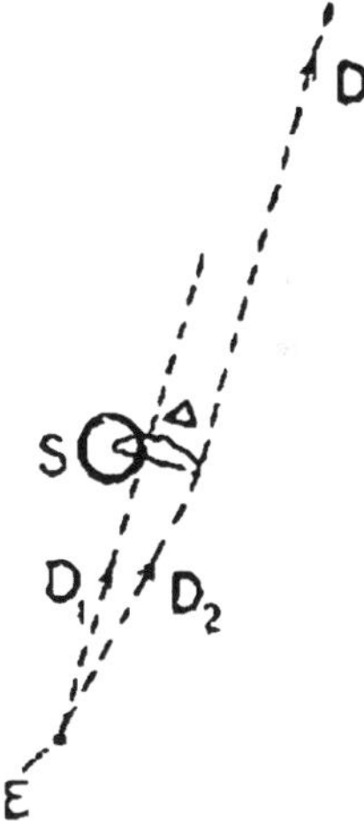

FIG. 5.

deflection of light from the star by the sun, the star will be seen in the direction D_2, *i.e.* at a somewhat greater distance from the centre of the sun than corresponds to its real position.

In practice, the question is tested in the following way. The stars in the neighbourhood of the sun are photographed during a solar eclipse. In addition, a second photograph of the same stars is taken when the sun is situated at another position in the sky, *i.e.* a few months earlier or later. As compared with the standard photograph, the positions of the stars on the eclipse-photograph ought to appear displaced radially outwards (away from the centre of the sun) by an amount corresponding to the angle α.

We are indebted to the Royal Society and to the Royal Astronomical Society for the investigation of this important deduction. Undaunted by the war and by difficulties of both a material and a psychological nature aroused by the war, these societies equipped two expeditions—to Sobral (Brazil), and to the island of Principe (West Africa)—and sent several of Britain's most celebrated astronomers (Eddington, Cottingham, Crommelin, Davidson), in order to obtain photographs of the solar eclipse of 29th May, 1919. The relative discrepancies to be expected between the stellar photographs obtained during the eclipse and the comparison photographs amounted to a few hundredths of a millimetre only. Thus great accuracy was necessary in making the adjustments required for the taking of the photographs, and in their subsequent measurement.

The results of the measurements confirmed the theory in a thoroughly satisfactory manner. The rectangular components of the observed and of the calculated deviations of the stars (in seconds of arc) are set forth in the following table of results :

Number of the Star.	First Co-ordinate.		Second Co-ordinate.	
	Observed.	Calculated.	Observed.	Calculated.
11 . .	−0·19	−0·22	+0·16	+0·02
5 . .	+0·29	+0·31	−0·46	−0·43
4 . .	+0·11	+0·10	+0·83	+0·74
3 . .	+0·20	+0·12	+1·00	+0·87
6 . .	+0·10	+0·04	+0·57	+0·40
10 . .	−0·08	+0·09	+0·35	+0·32
2 . .	+0·95	+0·85	−0·27	−0·09

(c) Displacement of Spectral Lines towards the Red

In Section XXIII it has been shown that in a system K' which is in rotation with regard to a Galileian system K, clocks of identical construction, and which are considered at rest with respect to the rotating reference-body, go at rates which are dependent on the positions of the clocks. We shall now examine this dependence quantitatively. A clock, which is situated at a distance r from the centre of the disc, has a velocity relative to K which is given by

$$v = \omega r,$$

where ω represents the angular velocity of rotation of the disc K' with respect to K. If ν_0 represents the number of ticks of the clock per unit time ("rate" of the clock) relative to K when the clock is at rest, then the "rate" of the clock (ν) when it is moving relative to K with a velocity v, but at rest with respect to the disc, will, in accordance with Section XII, be given by

$$\nu = \nu_0 \sqrt{1 - \frac{v^2}{c^2}},$$

or with sufficient accuracy by

$$\nu = \nu_0\left(1 - \frac{1}{2}\frac{v^2}{c^2}\right).$$

This expression may also be stated in the following form:

$$\nu = \nu_0\left(1 - \frac{1}{c^2}\frac{\omega^2 r^2}{2}\right).$$

If we represent the difference of potential of the centrifugal force between the position of the clock and the centre of the disc by ϕ, *i.e.* the work, considered negatively, which must be performed on the unit of mass against the centrifugal force in order to transport it from the position of the clock on the rotating disc to the centre of the disc, then we have

$$\phi = -\frac{\omega^2 r^2}{2}.$$

From this it follows that

$$\nu = \nu_0\left(1 + \frac{\phi}{c^2}\right).$$

In the first place, we see from this expression that two clocks of identical construction will go at different rates when situated at different distances from the centre of the disc. This result is also valid from the standpoint of an observer who is rotating with the disc.

Now, as judged from the disc, the latter is in a gravitational field of potential ϕ, hence the result we have obtained will hold quite generally for gravitational fields. Furthermore, we can regard an atom which is emitting spectral lines as a clock, so that the following statement will hold:

An atom absorbs or emits light of a frequency which is

dependent on the potential of the gravitational field in which it is situated.

The frequency of an atom situated on the surface of a heavenly body will be somewhat less than the frequency of an atom of the same element which is situated in free space (or on the surface of a smaller celestial body). Now $\phi = -K\frac{M}{r}$, where K is Newton's constant of gravitation, and M is the mass of the heavenly body. Thus a displacement towards the red ought to take place for spectral lines produced at the surface of stars as compared with the spectral lines of the same element produced at the surface of the earth, the amount of this displacement being

$$\frac{\nu_0 - \nu}{\nu_0} = \frac{K}{c^2}\frac{M}{r}.$$

For the sun, the displacement towards the red predicted by theory amounts to about two millionths of the wave-length. A trustworthy calculation is not possible in the case of the stars, because in general neither the mass M nor the radius r is known.

It is an open question whether or not this effect exists, and at the present time astronomers are working with great zeal towards the solution. Owing to the smallness of the effect in the case of the sun, it is difficult to form an opinion as to its existence. Whereas Grebe and Bachem (Bonn), as a result of their own measurements and those of Evershed and Schwarzschild on the cyanogen bands, have placed the existence of the effect almost beyond doubt, other investigators, particularly St. John, have been led to the opposite opinion in consequence of their measurements.

Mean displacements of lines towards the less refrangible end of the spectrum are certainly revealed by statistical investigations of the fixed stars; but up to the present the examination of the available data does not allow of any definite decision being arrived at, as to whether or not these displacements are to be referred in reality to the effect of gravitation. The results of observation have been collected together, and discussed in detail from the standpoint of the question which has been engaging our attention here, in a paper by E. Freundlich entitled "Zur Prüfung der allgemeinen Relativitäts-Theorie" (*Die Naturwissenschaften*, 1919, No. 35, p. 520: Julius Springer, Berlin).

At all events, a definite decision will be reached during the next few years. If the displacement of spectral lines towards the red by the gravitational potential does not exist, then the general theory of relativity will be untenable. On the other hand, if the cause of the displacement of spectral lines be definitely traced to the gravitational potential, then the study of this displacement will furnish us with important information as to the mass of the heavenly bodies.

BIBLIOGRAPHY

WORKS IN ENGLISH ON EINSTEIN'S THEORY

Introductory

The Foundations of Einstein's Theory of Gravitation: Erwin Freundlich (translation by H. L. Brose). Camb. Univ. Press, 1920.

Space and Time in Contemporary Physics: Moritz Schlick (translation by H. L. Brose). Clarendon Press, Oxford, 1920.

The Special Theory

The Principle of Relativity: E. Cunningham. Camb. Univ. Press.

Relativity and the Electron Theory: E. Cunningham, Monographs on Physics. Longmans, Green & Co.

The Theory of Relativity: L. Silberstein. Macmillan & Co.

The Space-Time Manifold of Relativity: E. B. Wilson and G. N. Lewis, *Proc. Amer. Soc. Arts & Science*, vol. xlviii., No. 11, 1912.

The General Theory

Report on the Relativity Theory of Gravitation: A. S. Eddington. Fleetway Press Ltd., Fleet Street, London.

On Einstein's Theory of Gravitation and its Astronomical Consequences: W. de Sitter, *M. N. Roy. Astron. Soc.*, lxxvi. p. 699, 1916; lxxvii. p. 155, 1916; lxxviii. p. 3, 1917.

On Einstein's Theory of Gravitation: H. A. Lorentz, *Proc. Amsterdam Acad.*, vol. xix. p. 1341, 1917.

Space, Time and Gravitation: W. de Sitter: *The Observatory*, No. 505, p. 412. Taylor & Francis, Fleet Street, London.

The Total Eclipse of 29th May, 1919, and the Influence of Gravitation on Light: A. S. Eddington, *ibid.*, March 1919.

Discussion on the Theory of Relativity: *M. N. Roy. Astron. Soc.*, vol. lxxx. No. 2, p. 96, December 1919.

The Displacement of Spectrum Lines and the Equivalence Hypothesis: W. G. Duffield, *M. N. Roy. Astron. Soc.*, vol. lxxx.; No. 3, p. 262, 1920.

Space, Time and Gravitation: A. S. Eddington, Camb. Univ. Press, 1920.

ALSO, CHAPTERS IN

The Mathematical Theory of Electricity and Magnetism: J. H. Jeans (4th edition). Camb. Univ. Press, 1920.

The Electron Theory of Matter: O. W. Richardson. Camb. Univ. Press.

INDEX

A SELECTION FROM

MESSRS. METHUEN'S PUBLICATIONS

This Catalogue contains only a selection of the more important books published by Messrs. Methuen. A complete catalogue of their publications may be obtained on application.

Bain (F. W.)—

A DIGIT OF THE MOON: A Hindoo Love Story. THE DESCENT OF THE SUN: A Cycle of Birth. A HEIFER OF THE DAWN. IN THE GREAT GOD'S HAIR. A DRAUGHT OF THE BLUE. AN ESSENCE OF THE DUSK. AN INCARNATION OF THE SNOW. A MINE OF FAULTS. THE ASHES OF A GOD. BUBBLES OF THE FOAM. A SYRUP OF THE BEES. THE LIVERY OF EVE. THE SUBSTANCE OF A DREAM. *All Fcap. 8vo.* 5*s. net.* AN ECHO OF THE SPHERES. *Wide Demy.* 12*s.* 6*d. net.*

Balfour (Graham). THE LIFE OF ROBERT LOUIS STEVENSON. *Fifteenth Edition. In one Volume. Cr. 8vo. Buckram,* 7*s.* 6*d. net.*

Belloc (H.)—

PARIS, 8*s.* 6*d. net.* HILLS AND THE SEA, 6*s. net.* ON NOTHING AND KINDRED SUBJECTS, 6*s. net.* ON EVERYTHING, 6*s. net.* ON SOMETHING, 6*s. net.* FIRST AND LAST, 6*s. net.* THIS AND THAT AND THE OTHER, 6*s. net.* MARIE ANTOINETTE, 18*s. net.* THE PYRENEES, 10*s.* 6*d. net.*

Bloemfontein (Bishop of). ARA CŒLI: AN ESSAY IN MYSTICAL THEOLOGY. *Seventh Edition. Cr. 8vo.* 5*s. net.*

FAITH AND EXPERIENCE. *Third Edition. Cr. 8vo.* 5*s. net.*

THE CULT OF THE PASSING MOMENT. *Fourth Edition. Cr. 8vo.* 5*s. net.*

THE ENGLISH CHURCH AND RE-UNION. *Cr. 8vo.* 5*s. net.*

SCALA MUNDI. *Cr. 8vo.* 4*s.* 6*d. net.*

Chesterton (G. K.)—

THE BALLAD OF THE WHITE HORSE. ALL THINGS CONSIDERED. TREMENDOUS TRIFLES. ALARMS AND DISCURSIONS. A MISCELLANY OF MEN. *All Fcap. 8vo.* 6*s. net.* WINE, WATER, AND SONG. *Fcap. 8vo.* 1*s.* 6*d net.*

Clutton-Brock (A.). WHAT IS THE KINGDOM OF HEAVEN? *Fourth Edition. Fcap. 8vo.* 5*s. net.*

ESSAYS ON ART. *Second Edition. Fcap. 8vo.* 5*s. net.*

Cole (G. D. H.). SOCIAL THEORY. *Cr. 8vo.* 5*s. net.*

Conrad (Joseph). THE MIRROR OF THE SEA: Memories and Impressions. *Fourth Edition. Fcap. 8vo.* 6*s. net.*

Einstein (A.). RELATIVITY: THE SPECIAL AND THE GENERAL THEORY. Translated by ROBERT W. LAWSON. *Cr. 8vo.* 5*s. net.*

Fyleman (Rose.). FAIRIES AND CHIMNEYS. *Fcap. 8vo. Sixth Edition.* 3*s.* 6*d. net.*

THE FAIRY GREEN. *Third Edition. Fcap. 8vo.* 3*s.* 6*d. net.*

Gibbins (H. de B.). INDUSTRY IN ENGLAND: HISTORICAL OUTLINES. With Maps and Plans. *Tenth Edition. Demy 8vo.* 12*s.* 6*d. net.*

THE INDUSTRIAL HISTORY OF ENGLAND. With 5 Maps and a Plan. *Twenty-seventh Edition. Cr. 8vo.* 5*s.*

Gibbon (Edward). THE DECLINE AND FALL OF THE ROMAN EMPIRE. Edited, with Notes, Appendices, and Maps, by J. B. BURY. Illustrated. *Seven Volumes. Demy 8vo.* Illustrated. *Each* 12*s.* 6*d. net. Also in Seven Volumes. Cr. 8vo. Each* 7*s.* 6*d. net.*

Glover (T. R.). THE CONFLICT OF RELIGIONS IN THE EARLY ROMAN EMPIRE. *Eighth Edition. Demy 8vo.* 10*s.* 6*d. net.*

POETS AND PURITANS. *Second Edition. Demy 8vo.* 10*s.* 6*d. net.*

FROM PERICLES TO PHILIP. *Third Edition. Demy 8vo.* 10*s.* 6*d. net.*

VIRGIL. *Fourth Edition. Demy 8vo.* 10*s.* 6*d. net.*

THE CHRISTIAN TRADITION AND ITS VERIFICATION. (The Angus Lecture for 1912.) *Second Edition. Cr. 8vo.* 6*s. net.*

Grahame (Kenneth). THE WIND IN THE WILLOWS. *Tenth Edition. Cr. 8vo.* 7*s.* 6*d. net.*

Hall (H. R.). THE ANCIENT HISTORY OF THE NEAR EAST FROM THE EARLIEST TIMES TO THE BATTLE OF SALAMIS. Illustrated. *Fourth Edition. Demy 8vo.* 16*s. net.*

Hobson (J. A.). INTERNATIONAL TRADE: AN APPLICATION OF ECONOMIC THEORY. *Cr. 8vo.* 5*s. net.*

PROBLEMS OF POVERTY: AN INQUIRY INTO THE INDUSTRIAL CONDITION OF THE POOR. *Eighth Edition. Cr. 8vo.* 5*s. net.*

THE PROBLEM OF THE UNEMPLOYED: AN INQUIRY AND AN ECONOMIC POLICY. *Sixth Edition. Cr. 8vo.* 5*s. net.*

GOLD, PRICES AND WAGES: WITH AN EXAMINATION OF THE QUANTITY THEORY. *Second Edition. Cr. 8vo.* 5s. *net.*

TAXATION IN THE NEW STATE. *Cr. 8vo.* 6s. *net.*

Holdsworth (W. S.). A HISTORY OF ENGLISH LAW. *Vol. I., II., III., Each Second Edition. Demy 8vo. Each* 15s. *net.*

Inge (W. R.). CHRISTIAN MYSTICISM. (The Bampton Lectures of 1899.) *Fourth Edition. Cr. 8vo.* 7s. 6d. *net.*

Jenks (E.). AN OUTLINE OF ENGLISH LOCAL GOVERNMENT. *Fourth Edition.* Revised by R. C. K. ENSOR. *Cr. 8vo.* 5s. *net.*

A SHORT HISTORY OF ENGLISH LAW: FROM THE EARLIEST TIMES TO THE END OF THE YEAR 1911. *Second Edition, revised. Demy 8vo.* 12s. 6d. *net.*

Julian (Lady) of Norwich. REVELATIONS OF DIVINE LOVE. Edited by GRACE WARRACK. *Seventh Edition. Cr. 8vo.* 5s. *net.*

Keats (John). POEMS. Edited, with Introduction and Notes, by E. de SÉLINCOURT. With a Frontispiece in Photogravure. *Third Edition. Demy 8vo.* 10s. 6d. *net.*

Kipling (Rudyard). BARRACK-ROOM BALLADS. 205*th Thousand. Cr. 8vo. Buckram,* 7s. 6d. *net. Also Fcap. 8vo. Cloth,* 6s. *net; leather,* 7s. 6d. *net.*
Also a Service Edition. *Two Volumes. Square fcap. 8vo. Each* 3s. *net.*

THE SEVEN SEAS. 152*nd Thousand. Cr. 8vo. Buckram,* 7s. 6d. *net. Also Fcap. 8vo. Cloth,* 6s. *net; leather,* 7s. 6d. *net.*
Also a Service Edition. *Two Volumes. Square fcap. 8vo. Each* 3s. *net.*

THE FIVE NATIONS. 126*th Thousand. Cr. 8vo. Buckram,* 7s. 6d. *net. Also Fcap. 8vo. Cloth,* 6s. *net; leather,* 7s. 6d. *net.*
Also a Service Edition. *Two Volumes. Square fcap. 8vo. Each* 3s. *net.*

DEPARTMENTAL DITTIES. 94*th Thousand. Cr. 8vo. Buckram,* 7s. 6d. *net. Also Fcap. 8vo. Cloth,* 6s. *net; leather,* 7s. 6d. *net.*
Also a Service Edition. *Two Volumes. Square fcap. 8vo. Each* 3s. *net.*

THE YEARS BETWEEN. *Cr. 8vo. Buckram,* 7s. 6d. *net. Also on thin paper. Fcap. 8vo. Blue cloth,* 6s. *net; Limp lambskin,* 7s. 6d. *net.*
Also a Service Edition. *Two Volumes. Square fcap. 8vo. Each* 3s. *net.*

HYMN BEFORE ACTION. Illuminated. *Fcap. 4to.* 1s. 6d. *net.*

RECESSIONAL. Illuminated. *Fcap. 4to.* 1s. 6d. *net.*

TWENTY POEMS FROM RUDYARD KIPLING. 360*th Thousand. Fcap. 8vo.* 1s. *net.*

Lamb (Charles and Mary). THE COMPLETE WORKS. Edited by E. V. LUCAS. *A New and Revised Edition in Six Volumes. With Frontispieces. Fcap. 8vo. Each* 6s. *net.*
The volumes are:—
I. MISCELLANEOUS PROSE. II. ELIA AND THE LAST ESSAY OF ELIA. III. BOOKS FOR CHILDREN. IV. PLAYS AND POEMS. V. and VI. LETTERS.

Lankester (Sir Ray). SCIENCE FROM AN EASY CHAIR. Illustrated. *Thirteenth Edition. Cr. 8vo.* 7s. 6d. *net.*

SCIENCE FROM AN EASY CHAIR. Illustrated. *Second Series. Third Edition. Cr. 8vo.* 7s. 6d. *net.*

DIVERSIONS OF A NATURALIST. Illustrated. *Third Edition. Cr. 8vo.* 7s. 6d. *net.*

SECRETS OF EARTH AND SEA. *Cr. 8vo.* 8s. 6d. *net.*

Lodge (Sir Oliver). MAN AND THE UNIVERSE: A STUDY OF THE INFLUENCE OF THE ADVANCE IN SCIENTIFIC KNOWLEDGE UPON OUR UNDERSTANDING OF CHRISTIANITY. *Ninth Edition. Crown 8vo.* 7s. 6d. *net.*

THE SURVIVAL OF MAN: A STUDY IN UNRECOGNISED HUMAN FACULTY. *Seventh Edition. Cr. 8vo.* 7s. 6d. *net.*

MODERN PROBLEMS. *Cr. 8vo.* 7s. 6d. *net.*

RAYMOND; OR LIFE AND DEATH. Illustrated. *Twelfth Edition. Demy 8vo.* 15s. *net.*

THE WAR AND AFTER: SHORT CHAPTERS ON SUBJECTS OF SERIOUS PRACTICAL IMPORT FOR THE AVERAGE CITIZEN IN A.D. 1915 ONWARDS. *Eighth Edition. Fcap 8vo.* 2s. *net.*

Lucas (E. V.).
THE LIFE OF CHARLES LAMB, 2 *vols.*, 21s. *net.* A WANDERER IN HOLLAND, 10s. 6d. *net.* A WANDERER IN LONDON, 10s. 6d. *net.* LONDON REVISITED, 10s. 6d. *net.* A WANDERER IN PARIS, 10s. 6d. *net* and 6s. *net.* A WANDERER IN FLORENCE, 10s. 6d. *net.* A WANDERER IN VENICE, 10s. 6d. *net.* THE OPEN ROAD: A Little Book for Wayfarers, 6s. 6d. *net* and 7s. 6d. *net.* THE FRIENDLY TOWN: A Little Book for the Urbane, 6s. *net.* FIRESIDE AND SUNSHINE, 6s. *net.* CHARACTER AND COMEDY, 6s. *net.* THE GENTLEST ART: A Choice of Letters by Entertaining Hands, 6s. 6d. *net.* THE SECOND POST, 6s. *net.* HER INFINITE VARIETY: A Feminine Portrait Gallery, 6s. *net.* GOOD COMPANY: A Rally of Men, 6s. *net.* ONE DAY AND ANOTHER, 6s. *net.* OLD LAMPS FOR NEW, 6s. *net.* LOITERER'S HARVEST, 6s. *net.* CLOUD AND SILVER, 6s. *net.* LISTENER'S LURE: An Oblique Narration, 6s. *net.* OVER BEMERTON'S: An Easy-Going Chronicle, 6s. *net.* MR. INGLESIDE, 6s. *net.* LONDON LAVENDER, 6s. *net.* LANDMARKS, 6s. *net.* A BOSWELL OF BAGHDAD, AND OTHER ESSAYS, 6s. *net.* 'TWIXT EAGLE AND DOVE, 6s. *net.* THE PHANTOM JOURNAL, AND OTHER ESSAYS AND DIVERSIONS, 6s. *net.* THE BRITISH SCHOOL: An Anecdotal Guide to the British Painters and Paintings in the National Gallery, 6s. *net.*

McDougall (William). AN INTRODUCTION TO SOCIAL PSYCHOLOGY. *Fifteenth Edition. Cr. 8vo. 7s. 6d. net.*

BODY AND MIND: A HISTORY AND A DEFENCE OF ANIMISM. *Fourth Edition. Demy 8vo. 12s. 6d. net.*

Maeterlinck (Maurice)—

THE BLUE BIRD: A Fairy Play in Six Acts, 6s. *net.* MARY MAGDALENE: A Play in Three Acts, 5s. *net.* DEATH, 3s. 6d. *net.* OUR ETERNITY, 6s. *net.* THE UNKNOWN GUEST, 6s. *net.* POEMS, 5s. *net.* THE WRACK OF THE STORM, 6s. *net.* THE MIRACLE OF ST. ANTHONY: A Play in One Act, 3s. 6d. *net.* THE BURGOMASTER OF STILEMONDE: A Play in Three Acts. 5s. *net.* THE BETROTHAL; or, The Blue Bird Chooses, 6s. *net.* MOUNTAIN PATHS, 6s. *net.*

Milne (A. A.). THE DAY'S PLAY. THE HOLIDAY ROUND. ONCE A WEEK. *All Cr. 8vo. 7s. net.* NOT THAT IT MATTERS. *Fcap 8vo. 6s. net.*

Oxenham (John)—

BEES IN AMBER: A Little Book of Thoughtful Verse. ALL'S WELL: A Collection of War Poems. THE KING'S HIGH WAY. THE VISION SPLENDID. THE FIERY CROSS. HIGH ALTARS: The Record of a Visit to the Battlefields of France and Flanders. HEARTS COURAGEOUS. ALL CLEAR! WINDS OF THE DAWN. *All Small Pott 8vo. Paper, 1s. 3d. net; cloth boards, 2s. net.* GENTLEMEN—THE KING, 2s. *net.*

Petrie (W. M. Flinders). A HISTORY OF EGYPT. Illustrated. *Six Volumes. Cr. 8vo. Each 9s. net.*

VOL. I. FROM THE IST TO THE XVITH DYNASTY. *Ninth Edition.* 10s 6d. *net.*

VOL. II. THE XVIITH AND XVIIITH DYNASTIES. *Sixth Edition.*

VOL. III. XIXTH TO XXXTH DYNASTIES. *Second Edition.*

VOL. IV. EGYPT UNDER THE PTOLEMAIC DYNASTY. J. P. MAHAFFY. *Second Edition.*

VOL. V. EGYPT UNDER ROMAN RULE. J. G. MILNE. *Second Edition.*

VOL. VI. EGYPT IN THE MIDDLE AGES. STANLEY LANE POOLE. *Second Edition.*

SYRIA AND EGYPT, FROM THE TELL EL AMARNA LETTERS. *Cr. 8vo. 5s. net.*

EGYPTIAN TALES. Translated from the Papyri. First Series, IVth to XIIth Dynasty. Illustrated. *Third Edition. Cr. 8vo. 5s. net.*

EGYPTIAN TALES. Translated from the Papyri. Second Series, XVIIITH to XIXTH Dynasty. Illustrated. *Second Edition. Cr. 8vo. 5s. net.*

Pollard (A. F.). A SHORT HISTORY OF THE GREAT WAR. With 19 Maps. *Second Edition. Cr. 8vo. 10s. 6d. net.*

Price (L. L.). A SHORT HISTORY OF POLITICAL ECONOMY IN ENGLAND FROM ADAM SMITH TO ARNOLD TOYNBEE. *Ninth Edition. Cr. 8vo. 5s. net.*

Reid (G. Archdall). THE LAWS OF HEREDITY. *Second Edition. Demy 8vo. £1 1s. net.*

Robertson (C. Grant). SELECT STATUTES, CASES, AND DOCUMENTS, 1660–1832. *Third Edition. Demy 8vo. 15s. net.*

Selous (Edmund). TOMMY SMITH'S ANIMALS. Illustrated. *Eighteenth Edition. Fcap. 8vo. 3s. 6d. net.*

TOMMY SMITH'S OTHER ANIMALS. Illustrated. *Eleventh Edition. Fcap. 8vo. 3s. 6d. net.*

TOMMY SMITH AT THE ZOO. Illustrated. *Fourth Edition. Fcap. 8vo. 2s. 9d.*

TOMMY SMITH AGAIN AT THE ZOO. Illustrated. *Second Edition. Fcap. 8vo. 2s. 9d.*

JACK'S INSECTS. Illustrated. *Cr. 8vo. 6s. net.*

JACK'S INSECTS. *Popular Edition. Vol. I. Cr. 8vo. 3s. 6d.*

Shelley (Percy Bysshe). POEMS. With an Introduction by A. CLUTTON-BROCK and Notes by C. D. LOCOCK. *Two Volumes. Demy 8vo. £1 1s. net.*

Smith (Adam). THE WEALTH OF NATIONS. Edited by EDWIN CANNAN. *Two Volumes. Second Edition. Demy 8vo. £1 5s. net.*

Stevenson (R. L.). THE LETTERS OF ROBERT LOUIS STEVENSON. Edited by Sir SIDNEY COLVIN. *A New Rearranged Edition in four volumes. Fourth Edition. Fcap. 8vo. Each 6s. net.*

Surtees (R. S.). HANDLEY CROSS. Illustrated. *Ninth Edition. Fcap. 8vo. 7s. 6d. net.*

MR. SPONGE'S SPORTING TOUR. Illustrated. *Fifth Edition. Fcap. 8vo. 7s. 6d. net.*

ASK MAMMA: OR, THE RICHEST COMMONER IN ENGLAND. Illustrated. *Second Edition. Fcap. 8vo. 7s. 6d. net.*

JORROCKS'S JAUNTS AND JOLLITIES. Illustrated. *Seventh Edition. Fcap. 8vo. 6s. net.*

MR. FACEY ROMFORD'S HOUNDS. Illustrated. *Third Edition. Fcap. 8vo. 7s. 6d. net.*

HAWBUCK GRANGE; OR, THE SPORTING ADVENTURES OF THOMAS SCOTT, ESQ. Illustrated. *Fcap. 8vo. 6s. net.*

PLAIN OR RINGLETS? Illustrated. *Fcap. 8vo. 7s. 6d. net.*

HILLINGDON HALL. With 12 Coloured Plates by WILDRAKE, HEATH, and JELLICOE. *Fcap. 8vo. 7s. 6d. net.*

Tileston (Mary W.). DAILY STRENGTH FOR DAILY NEEDS. *Twenty-sixth Edition. Medium 16mo.* 3s. 6d. *net.*

Underhill (Evelyn). MYSTICISM. A Study in the Nature and Development of Man's Spiritual Consciousness. *Eighth Edition. Demy 8vo.* 15s. *net.*

Vardon (Harry). HOW TO PLAY GOLF. Illustrated. *Thirteenth Edition. Cr. 8vo.* 5s. *net.*

Waterhouse (Elizabeth). A LITTLE BOOK OF LIFE AND DEATH. *Twentieth Edition. Small Pott 8vo. Cloth,* 2s. 6d. *net.*

Wells (J.). A SHORT HISTORY OF ROME. *Seventeenth Edition.* With 3 Maps. *Cr. 8vo.* 6s.

Wilde (Oscar). THE WORKS OF OSCAR WILDE. *Fcap. 8vo. Each* 6s. 6d. *net.*

I. LORD ARTHUR SAVILE'S CRIME AND THE PORTRAIT OF MR. W. H. II. THE DUCHESS OF PADUA. III. POEMS. IV. LADY WINDERMERE'S FAN. V. A WOMAN OF NO IMPORTANCE. VI. AN IDEAL HUSBAND. VII. THE IMPORTANCE OF BEING EARNEST. VIII. A HOUSE OF POMEGRANATES. IX. INTENTIONS. X. DE PROFUNDIS AND PRISON LETTERS. XI. ESSAYS. XII. SALOMÉ, A FLORENTINE TRAGEDY, and LA SAINTE COURTISANE. XIII. A CRITIC IN PALL MALL. XIV. SELECTED PROSE OF OSCAR WILDE. XV. ART AND DECORATION.

A HOUSE OF POMEGRANATES. Illustrated. *Cr. 4to.* 21s. *net.*

Wood (Lieut. W. B.) and Edmonds (Col. J. E.). A HISTORY OF THE CIVIL WAR IN THE UNITED STATES (1861-65). With an Introduction by SPENSER WILKINSON. With 24 Maps and Plans. *Third Edition. Demy 8vo.* 15s. *net.*

Wordsworth (W.). POEMS. With an Introduction and Notes by NOWELL C. SMITH. *Three Volumes. Demy 8vo.* 18s. *net.*

Yeats (W. B.). A BOOK OF IRISH VERSE. *Fourth Edition. Cr. 8vo.* 7s. *net.*

PART II.—A SELECTION OF SERIES

Ancient Cities

General Editor, SIR B. C. A. WINDLE

Cr. 8vo. 6s. *net each volume*

With Illustrations by E. H. NEW, and other Artists

BRISTOL. CANTERBURY. CHESTER. DUBLIN. EDINBURGH. LINCOLN. SHREWSBURY. WELLS and GLASTONBURY.

The Antiquary's Books

General Editor, J. CHARLES COX

Demy 8vo. 10s. 6d. *net each volume*

With Numerous Illustrations

ANCIENT PAINTED GLASS IN ENGLAND. ARCHÆOLOGY AND FALSE ANTIQUITIES. THE BELLS OF ENGLAND. THE BRASSES OF ENGLAND. THE CASTLES AND WALLED TOWNS OF ENGLAND. CELTIC ART IN PAGAN AND CHRISTIAN TIMES. CHURCHWARDENS' ACCOUNTS. THE DOMESDAY INQUEST. ENGLISH CHURCH FURNITURE. ENGLISH COSTUME. ENGLISH MONASTIC LIFE. ENGLISH SEALS. FOLK-LORE AS AN HISTORICAL SCIENCE. THE GILDS AND COMPANIES OF LONDON. THE HERMITS AND ANCHORITES OF ENGLAND. THE MANOR AND MANORIAL RECORDS. THE MEDIÆVAL HOSPITALS OF ENGLAND. OLD ENGLISH INSTRUMENTS OF MUSIC. OLD ENGLISH LIBRARIES. OLD SERVICE BOOKS OF THE ENGLISH CHURCH. PARISH LIFE IN MEDIÆVAL ENGLAND. THE PARISH REGISTERS OF ENGLAND. REMAINS OF THE PREHISTORIC AGE IN ENGLAND. THE ROMAN ERA IN BRITAIN. ROMANO-BRITISH BUILDINGS AND EARTHWORKS. THE ROYAL FORESTS OF ENGLAND. THE SCHOOLS OF MEDIEVAL ENGLAND. SHRINES OF BRITISH SAINTS.

The Arden Shakespeare

General Editor, R. H. CASE

Demy 8vo. 6s. net each volume

An edition of Shakespeare in Single Plays; each edited with a full Introduction, Textual Notes, and a Commentary at the foot of the page.

Classics of Art

Edited by DR. J. H. W. LAING

With numerous Illustrations. Wide Royal 8vo

THE ART OF THE GREEKS, 15s. *net*. THE ART OF THE ROMANS, 16s. *net*. CHARDIN, 15s. *net*. DONATELLO, 16s. *net*. GEORGE ROMNEY, 15s. *net*. GHIRLANDAIO, 15s. *net*. LAWRENCE, 25s. *net*. MICHELANGELO, 15s. *net*. RAPHAEL, 15s. *net*. REMBRANDT'S ETCHINGS, Two Vols., 25s. *net*. TINTORETTO, 16s. *net*. TITIAN, 16s. *net*. TURNER'S SKETCHES AND DRAWINGS, 15s. *net*. VELAZQUEZ, 15s. *net*.

The 'Complete' Series

Fully Illustrated. Demy 8vo

THE COMPLETE AMATEUR BOXER, 10s. 6*d*. *net*. THE COMPLETE ASSOCIATION FOOTBALLER, 10s. 6*d*. *net*. THE COMPLETE ATHLETIC TRAINER, 10s. 6*d*. *net*. THE COMPLETE BILLIARD PLAYER, 12s. 6*d*. *net*. THE COMPLETE COOK, 10s. 6*d*. *net*. THE COMPLETE CRICKETER, 10s. 6*d*. *net*. THE COMPLETE FOXHUNTER, 16s. *net*. THE COMPLETE GOLFER, 12s. 6*d*. *net*. THE COMPLETE HOCKEY-PLAYER, 10s. 6*d*. *net*. THE COMPLETE HORSEMAN, 12s. 6*d*. *net*. THE COMPLETE JUJITSUAN, 5s. *net*. THE COMPLETE LAWN TENNIS PLAYER, 12s. 6*d*. *net*. THE COMPLETE MOTORIST, 10s. 6*d*. *net*. THE COMPLETE MOUNTAINEER, 16s. *net*. THE COMPLETE OARSMAN, 15s. *net*. THE COMPLETE PHOTOGRAPHER, 15s. *net*. THE COMPLETE RUGBY FOOTBALLER, ON THE NEW ZEALAND SYSTEM, 12s. 6*d*. *net*. THE COMPLETE SHOT, 16s. *net*. THE COMPLETE SWIMMER, 10s. 6*d*. *net*. THE COMPLETE YACHTSMAN, 16s. *net*.

The Connoisseur's Library

With numerous Illustrations. Wide Royal 8vo. 25s. net each volume

ENGLISH COLOURED BOOKS. ENGLISH FURNITURE. ETCHINGS. EUROPEAN ENAMELS. FINE BOOKS. GLASS. GOLDSMITHS' AND SILVERSMITHS' WORK. ILLUMINATED MANUSCRIPTS. IVORIES. JEWELLERY. MEZZOTINTS. MINIATURES. PORCELAIN. SEALS. WOOD SCULPTURE.

Handbooks of Theology

Demy 8vo

THE DOCTRINE OF THE INCARNATION, 15s. *net*. A HISTORY OF EARLY CHRISTIAN DOCTRINE, 16s. *net*. INTRODUCTION TO THE HISTORY OF RELIGION, 12s. 6*d*. *net*. AN INTRODUCTION TO THE HISTORY OF THE CREEDS, 12s. 6*d*. *net*. THE PHILOSOPHY OF RELIGION IN ENGLAND AND AMERICA, 12s. 6*d*. *net*. THE XXXIX ARTICLES OF THE CHURCH OF ENGLAND, 15s. *net*.

Health Series

Fcap. 8vo. 2s. 6d. net

THE BABY. THE CARE OF THE BODY. THE CARE OF THE TEETH. THE EYES OF OUR CHILDREN. HEALTH FOR THE MIDDLE-AGED. THE HEALTH OF A WOMAN. THE HEALTH OF THE SKIN. HOW TO LIVE LONG. THE PREVENTION OF THE COMMON COLD. STAYING THE PLAGUE. THROAT AND EAR TROUBLES. TUBERCULOSIS. THE HEALTH OF THE CHILD, 2s. *net*.

Leaders of Religion

Edited by H. C. BEECHING. *With Portraits*

Crown 8vo. 3s. net each volume

The Library of Devotion

Handy Editions of the great Devotional Books, well edited. With Introductions and (where necessary) Notes

Small Pott 8vo, cloth, 3s. net and 3s. 6d. net

Little Books on Art

With many Illustrations. Demy 16mo. 5s. net each volume

Each volume consists of about 200 pages, and contains from 30 to 40 Illustrations, including a Frontispiece in Photogravure

ALBRECHT DÜRER. THE ARTS OF JAPAN. BOOKPLATES. BOTTICELLI. BURNE-JONES. CELLINI. CHRISTIAN SYMBOLISM. CHRIST IN ART. CLAUDE. CONSTABLE. COROT. EARLY ENGLISH WATER-COLOUR. ENAMELS. FREDERIC LEIGHTON. GEORGE ROMNEY. GREEK ART. GREUZE AND BOUCHER. HOLBEIN. ILLUMINATED MANUSCRIPTS. JEWELLERY. JOHN HOPPNER. SIR JOSHUA REYNOLDS. MILLET. MINIATURES. OUR LADY IN ART. RAPHAEL. RODIN. TURNER. VANDYCK. VELAZQUEZ. WATTS.

The Little Guides

With many Illustrations by E. H. NEW and other artists, and from photographs

Small Pott 8vo. 4s. net and 6s. net

Guides to the English and Welsh Counties, and some well-known districts

The main features of these Guides are (1) a handy and charming form; (2) illustrations from photographs and by well-known artists; (3) good plans and maps; (4) an adequate but compact presentation of everything that is interesting in the natural features, history, archæology, and architecture of the town or district treated.

The Little Quarto Shakespeare

Edited by W. J. CRAIG. With Introductions and Notes

Pott 16mo. 40 Volumes. Leather, price 1s. 9d. net each volume

Cloth, 1s. 6d.

Nine Plays

Fcap. 8vo. 3s. 6d. net

ACROSS THE BORDER. Beulah Marie Dix. *Cr. 8vo.*

HONEYMOON, THE. A Comedy in Three Acts. Arnold Bennett. *Third Edition.*

GREAT ADVENTURE, THE. A Play of Fancy in Four Acts. Arnold Bennett. *Fifth Edition.*

MILESTONES. Arnold Bennett and Edward Knoblock. *Ninth Edition.*

IDEAL HUSBAND, AN. Oscar Wilde. *Acting Edition.*

KISMET. Edward Knoblock. *Fourth Edition.*

TYPHOON. A Play in Four Acts. Melchior Lengyel. English Version by Laurence Irving. *Second Edition.*

WARE CASE, THE. George Pleydell.

GENERAL POST. J. E. Harold Terry. *Second Edition.*

Sports Series

Illustrated. Fcap. 8vo. 2s. net and 3s. net

ALL ABOUT FLYING, 3s. *net.* GOLF DO'S AND DONT'S. THE GOLFING SWING. HOW TO SWIM. LAWN TENNIS, 3s. *net.* SKATING, 3s. *net.* CROSS-COUNTRY SKI-ING, 5s. *net.* WRESTLING, 2s. *net.* QUICK CUTS TO GOOD GOLF, 2s. 6d. *net.*

The Westminster Commentaries

General Editor, WALTER LOCK

Demy 8vo

THE ACTS OF THE APOSTLES, 16s. *net.* AMOS, 8s. 6d. *net.* I. CORINTHIANS, 8s. 6d. *net.* EXODUS, 15s. *net.* EZEKIEL, 12s. 6d. *net.* GENESIS, 16s. *net.* HEBREWS, 8s. 6d. *net.* ISAIAH, 16s. *net.* JEREMIAH, 16s. *net.* JOB, 8s. 6d. *net.* THE PASTORAL EPISTLES, 8s. 6d. *net.* THE PHILIPPIANS, 8s. 6d. *net.* ST. JAMES, 8s. 6d. *net.* ST. MATTHEW, 15s. *net.*

Methuen's Two-Shilling Library

Cheap Editions of many Popular Books

Fcap. 8vo

PART III.—A SELECTION OF WORKS OF FICTION

Bennett (Arnold)—

CLAYHANGER, 8s. *net.* HILDA LESSWAYS, 8s. 6d. *net.* THESE TWAIN. THE CARD. THE REGENT: A Five Towns Story of Adventure in London. THE PRICE OF LOVE. BURIED ALIVE. A MAN FROM THE NORTH. THE MATADOR OF THE FIVE TOWNS. WHOM GOD HATH JOINED. A GREAT MAN: A Frolic. *All 7s. 6d. net.*

Birmingham (George A.)—

SPANISH GOLD. THE SEARCH PARTY. LALAGE'S LOVERS. THE BAD TIMES. UP, THE REBELS. *All 7s. 6d. net.*

Burroughs (Edgar Rice)—

TARZAN OF THE APES, 6s. *net.* THE RETURN OF TARZAN, 6s. *net.* THE BEASTS OF TARZAN, 6s. *net.* THE SON OF TARZAN, 6s. *net.* JUNGLE TALES OF TARZAN, 6s. *net.* TARZAN AND THE JEWELS OF OPAR, 6s. *net.* TARZAN THE UNTAMED, 7s. 6d. *net.* A PRINCESS OF MARS, 6s. *net.* THE GODS OF MARS, 6s. *net.* THE WARLORD OF MARS, 6s. *net.*

Conrad (Joseph). A SET OF SIX. *Fourth Edition. Cr. 8vo.* 7s. 6d. *net.*

VICTORY: AN ISLAND TALE. *Sixth Edition. Cr. 8vo.* 9s. *net.*

Corelli (Marie)—

A ROMANCE OF TWO WORLDS, 7s. 6d. *net.* VENDETTA: or, The Story of One Forgotten, 8s. *net.* THELMA: A Norwegian Princess, 8s. 6d. *net.* ARDATH: The Story of a Dead Self, 7s. 6d. *net.* THE SOUL OF LILITH, 7s. 6d. *net.* WORMWOOD: A Drama of Paris, 8s. *net.* BARABBAS: A Dream of the World's Tragedy, 8s. *net.* THE SORROWS OF SATAN, 7s. 6d. *net.* THE MASTER-CHRISTIAN, 8s. 6d. *net.* TEMPORAL POWER: A Study in Supremacy, 6s. *net.* GOD'S GOOD MAN: A Simple Love Story, 8s. 6d. *net.* HOLY ORDERS: The Tragedy of a Quiet Life, 8s. 6d. *net.* THE MIGHTY ATOM, 7s. 6d. *net.* BOY: A Sketch, 7s. 6d. *net.* CAMEOS, 6s. *net.* THE LIFE EVERLASTING, 8s. 6d. *net.*

Doyle (Sir A. Conan). ROUND THE RED LAMP. *Twelfth Edition. Cr. 8vo.* 7s. 6d. *net.*

Hichens (Robert)—

TONGUES OF CONSCIENCE, 7s. 6d. *net.* FELIX: Three Years in a Life, 7s. 6d. *net.* THE WOMAN WITH THE FAN, 7s. 6d. *net.* BYEWAYS, 7s. 6d. *net.* THE GARDEN OF ALLAH, 8s. 6d. *net.* THE CALL OF THE BLOOD, 8s. 6d. *net.* BARBARY SHEEP, 6s. *net.* THE DWELLERS ON THE THRESHOLD, 7s. 6d. *net.* THE WAY OF AMBITION, 7s. 6d. *net.* IN THE WILDERNESS, 7s. 6d. *net.*

Hope (Anthony)—
A CHANGE OF AIR. A MAN OF MARK. THE CHRONICLES OF COUNT ANTONIO. SIMON DALE. THE KING'S MIRROR. QUISANTÉ. THE DOLLY DIALOGUES. TALES OF TWO PEOPLE. A SERVANT OF THE PUBLIC. MRS. MAXON PROTESTS. A YOUNG MAN'S YEAR. BEAUMAROY HOME FROM THE WARS. *All 7s. 6d. net.*

Jacobs (W. W.)—
MANY CARGOES, *5s. net* and *2s. 6d. net.* SEA URCHINS, *5s. net* and *3s. 6d. net.* A MASTER OF CRAFT, *5s. net.* LIGHT FREIGHTS, *5s. net.* THE SKIPPER'S WOOING, *5s. net.* AT SUNWICH PORT, *5s. net.* DIALSTONE LANE, *5s. net.* ODD CRAFT, *5s. net.* THE LADY OF THE BARGE, *5s. net.* SALTHAVEN, *5s. net.* SAILORS' KNOTS, *5s. net.* SHORT CRUISES, *5s. net.*

London (Jack). WHITE FANG. *Ninth Edition. Cr. 8vo. 7s. 6d. net.*

McKenna (Stephen)—
SONIA: Between Two Worlds, *8s. net.* NINETY-SIX HOURS' LEAVE, *7s. 6d. net.* THE SIXTH SENSE, *6s. net.* MIDAS & SON, *8s. net.*

Malet (Lucas)—
THE HISTORY OF SIR RICHARD CALMADY: A Romance. THE WAGES OF SIN. THE CARISSIMA. THE GATELESS BARRIER. DEADHAM HARD. *All 7s. 6d. net.*

Mason (A. E. W.). CLEMENTINA. Illustrated. *Ninth Edition. Cr. 8vo. 7s. 6d. net.*

Maxwell (W. B.)—
VIVIEN. THE GUARDED FLAME. ODD LENGTHS. HILL RISE. THE REST CURE. *All 7s. 6d. net.*

Oxenham (John)—
A WEAVER OF WEBS. PROFIT AND LOSS. THE SONG OF HYACINTH, and Other Stories. LAURISTONS. THE COIL OF CARNE. THE QUEST OF THE GOLDEN ROSE. MARY ALL-ALONE. BROKEN SHACKLES. "1914." *All 7s. 6d. net.*

Parker (Gilbert)—
PIERRE AND HIS PEOPLE. MRS. FALCHION. THE TRANSLATION OF A SAVAGE. WHEN VALMOND CAME TO PONTIAC: The Story of a Lost Napoleon. AN ADVENTURER OF THE NORTH: The Last Adventures of 'Pretty Pierre.' THE SEATS OF THE MIGHTY. THE BATTLE OF THE STRONG: A Romance of Two Kingdoms. THE POMP OF THE LAVILETTES. NORTHERN LIGHTS. *All 7s. 6d. net.*

Phillpotts (Eden)—
CHILDREN OF THE MIST. SONS OF THE MORNING. T[illegible] RIVER. THE AMERICAN PRISONER. [illegible]ER'S DAUGHTER. THE HUMAN BOY [illegible]HE WAR. *All 7s. 6d. net.*

Ridge (W. Pett)—
A SON OF THE STATE, *7s. 6d. net.* THE REMINGTON SENTENCE, *7s. 6d. net.* MADAME PRINCE, *7s. 6d. net.* TOP SPEED, *7s. 6d. net.* SPECIAL PERFORMANCES, *6s. net.* THE BUSTLING HOURS, *7s. 6d. net.*

Rohmer (Sax)—
THE DEVIL DOCTOR. THE SI-FAN MYSTERIES. TALES OF SECRET EGYPT. THE ORCHARD OF TEARS. THE GOLDEN SCORPION. *All 7s. 6d. net.*

Swinnerton (F.). SHOPS AND HOUSES. *Third Edition. Cr. 8vo. 7s. 6d. net.* SEPTEMBER. *Third Edition. Cr. 8vo. 7s. 6d. net.*

Wells (H. G.). BEALBY. *Fourth Edition. Cr. 8vo. 7s. 6d. net.*

Williamson (C. N. and A. M.)—
THE LIGHTNING CONDUCTOR: The Strange Adventures of a Motor Car. LADY BETTY ACROSS THE WATER. SCARLET RUNNER. LORD LOVELAND DISCOVERS AMERICA. THE GUESTS OF HERCULES. IT HAPPENED IN EGYPT. A SOLDIER OF THE LEGION. THE SHOP GIRL. THE LIGHTNING CONDUCTRESS. SECRET HISTORY. THE LOVE PIRATE. *All 7s. 6d. net.* CRUCIFIX CORNER. *6s. net.*

Methuen's Two-Shilling Novels

Cheap Editions of many of the most Popular Novels of the day

Write for Complete List

Fcap. 8vo

Made in the USA
Lexington, KY
07 March 2011